Acoustic Emission

Acoustic Emission

R V Williams
British Steel Corporation (Overseas Services) Ltd

Adam Hilger Ltd, Bristol

British Library Cataloguing in Publication Data

Williams, R V
  Acoustic emission.
  1. Acoustic emission
  I. Title
  620.1'1294   TA418.84

  ISBN 0-85274-359-9

Published by Adam Hilger Ltd,
Techno House, Redcliffe Way, Bristol BS1 6NX.

The Adam Hilger book-publishing imprint is owned by
The Institute of Physics.

Filmset and printed in Great Britain by Page Brothers
(Norwich) Ltd, Mile Cross Lane, Norwich NR6 6SA.

To my teachers

Contents

Preface

The loss of an accommodation platform in the North Sea early in 1980 highlighted the need for improved safety checks on the integrity of modern highly stressed structures. Engineers have always attempted to design equipment to take full advantage of the materials available to them, and developments in the aircraft, nuclear and offshore industries have in recent years pushed steels and aluminium alloys to the limits of resistance to stress and fatigue. As a consequence, even when full advantage can be taken of large-scale testing, the number of structures placed at risk due to material failure is greater than in the earlier days of cautious over-design. Fracture mechanics and advances in the metallurgist's understanding of material failure mechanisms have enabled designers to devise structures inconceivable even ten years ago. Economic and technical factors dictate that modern materials are used under conditions where failure can occur unless construction methods are nearly perfect. At the same time the economic and social effects of failure in, for example, a nuclear power station or an oil rig are very great. There is a clear need for non-destructive testing of such structures during construction and service, and acoustic emission techniques can now provide safety checks for most modern highly stressed equipment. Moreover, such checks can be carried out in service, twenty four hours a day under the severest conditions. This recent development has not hitherto been reviewed at length and so I have set out a dispassionate account of the capabilities of this technique in a way which will appeal to those seeking guidance on possible new applications, as well as to the specialist. Acoustic emission is not in itself the complete answer to integrity monitoring under critical stress conditions — conventional methods can also play their part — but it is the most powerful of the modern methods available to those concerned with the safety of structures.

I am fortunate in having been involved throughout the development work, carried out over the past five years, leading to the successful application of acoustic emission to the integrity and safety inspection of offshore, as well as many land-based, structures. I have tried to give details of both the acoustic emission equipment and its practical applications. Due reference is made to the now mature science of fracture mechanics: somewhat less detailed attention is given to purely metallurgical factors, but reference is made to papers — some highly theoretical — discussing microstructural phenomena and their relation to acoustic emission.

The acoustic emission community achieves rapid and full exchange of information between its members. This book is a distillation of the information available from many sources: various publications, conference notes and, in some cases, as yet unpublished papers. I have found the acoustic emission community particularly generous in its help and guidance. The over-enthusiastic claims of the early days of the technique have been replaced by a healthy, sceptical approach to new applications, backed by testing methods leading to real success in application in the field. All this I have tried to set down in this monograph. Needless to say any errors of fact or judgement are my own.

Dr R V Williams
21 April 1980

Acknowledgments

I would like to thank the many members of the acoustic emission fraternity who have helped me to understand the art and science of acoustic emission. I am particularly grateful to my colleagues in the Unit Inspection Company, who have generously given of their time and results while this book was in preparation. Viv Peters and Len Rogers were more than helpful; the chapter on offshore applications is taken almost entirely from their work. Dr Eric Duckworth of the Fulmer Research Institute generously provided material for the section on the testing of concrete, and Don Birchon, the father of acoustic emission in the United Kingdom, was most helpful to me when I was new to the field.

Acoustic emission is a particularly fortunate discipline in that it possesses a vigorous fraternity of practitioners in EWGAE — the European Working Group on Acoustic Emission. Members of this club have been generous in their help and advice, particularly Peter Bartle of the Welding Institute, Manlio Mirabile of Centro Sperimentale Metallurgico, Rome, and Aved Nielsen of the Risø National Laboratory, Rosskilde, Denmark. All three contributed to my knowledge of this interesting but sometimes difficult subject, and have patiently explained its finer points. Their understanding of acoustic emission is immense, and the science of non-destructive testing owes much to them. It goes without saying that any errors in this book are entirely due to the author.

Two talented young ladies, Mrs Freye Kennedy and Miss Jenny Wood, have successfully interpreted my indecipherable handwriting with patience and tact to produce an excellent text which has been turned into a fine book by the efforts of Paul Nagle and the other staff of Adam Hilger Ltd. Lastly, a word of thanks to my family who have lived with the writing of this book for some time with their usual good-natured tolerance.

1 Introduction

It has been known for many years that when a solid is subjected to stress at certain levels, discrete acoustic wave packets are generated which can be detected by transducers placed on, or in acoustic contact with, the solid. The phenomenon of sound generation in materials under stress is termed acoustic emission (AE), or, alternatively, stress wave emission. The purpose of the present monograph is to discuss this phenomenon with particular reference to its application as a non-destructive testing (NDT) technique. The monograph also includes a considerable section devoted to the design and use of systems able to detect the very low levels of acoustic energy generated by acoustic emission processes. Acoustic emission can be explained in terms of dislocation and other deformation processes in materials but these are not discussed in depth; rather the applications of the technique in construction and engineering are more extensively discussed.

Most materials which are designed to withstand high stress levels emit acoustic energy when stressed, including the well known metallic alloys such as steels, cast irons and alloys of aluminium. Glasses and fibres (as the high yield stress component of composite structures) as well as concrete and ceramic materials also emit acoustic energy when stressed. In view of the importance of concrete structures some attention is paid to the use of acoustic emission for the testing of concrete beams.

The subject is now of some reasonable antiquity. The noise due to the twinning experienced by fine tin when deformed is described in elementary chemistry textbooks and is known as the 'cry of tin'. As early as 1938, Frenkel and Kontorova proposed that the process of plastic deformation or twinning consists of a caterpillar-like motion of one atomic chain over another. These authors predicted acoustic emissions as did Frank (1949) in another early publication. Modern equipment can detect emissions produced by very small strain levels, as low as 0·05% or less in certain materials, and, approaching fracture, emissions can be very easily detected from most engineering materials.

This monograph was written because there appears to be no coherent account of the state of acoustic emission technology in the late 1970s. Excellent earlier publications do exist, notably the two sets of conference proceedings, published by the American Society for Testing and Materials, *Special Technical Publications 505* (1972) and *571* (1975) (these will be frequently referred to in the text as STP 505 and STP 571). Also, Dr R W Nichols (1976) has edited a series of papers, presented at a conference in Israel, in a book entitled *Acoustic Emission*. The third major publication in this field is a review article written by Ying (1973) in the *CRC Critical Reviews in Solid State Sciences*. This concentrates on the solid state physics aspects of the subject with rather less emphasis on applications; it is commended to the reader's attention. Also well worth noting is a bibliography compiled by Drolliard (1979).

Mention must be made immediately of the phenomenon observed in acoustic emission known as the Kaiser effect. This was discovered during PhD studies by Kaiser in Germany and published in his thesis from the Technische Hochschule Munich in 1950; an account is also given in Nichols (1976). Kaiser observed acoustic emission from polycrystallised specimens of zinc, steel, aluminium, copper and lead. While suggesting that grain boundary sliding was a source of the emission, he also noted that if crystals are stressed while emissions are being monitored and the stress is then relaxed, no new emissions will occur until the previous maximum stress has been exceeded. This phenomenon is not universal; it is associated with the incidence of the Bauschinger effect in the material under observation (the Bauschinger effect refers to the decrease in compression yield strength and increase in tensile yield when a metal is plastically strained beyond yield). Reference to the Kaiser effect is made in several later chapters of the present monograph.

It is appropriate at this stage, before going on to study applications and systems in more detail, to quickly review the advantages and disadvantages and state of application of the technique at the time of writing. Acoustic emission is essentially a method of non-destructively testing engineering structures which have to be examined for quality assurance or safety reasons. It responds, because of the Kaiser effect, only to stress levels which are greater than those which have been previously, and in many cases safely, maintained.

Arguably the most important reason why the method is now being actively used for testing structures is that it is essentially non-localised; it is not necessary for the receiving transducer to be particularly near the source of the emissions or the area which is under test. For example, a critical node in an offshore structure could be tested in service by only one dozen transducers while ultrasonic methods would need about a thousand sensors. Welds can be also checked during welding by just one transducer on the welding head. It is very easy to scan a large structure using acoustic emission probes placed at 1–10 m intervals on the surface of a structure. This ability to examine a large volume of material is in direct contrast to alternative methods of non-destructive examination such as ultrasonics or radiography. Ultrasonics in particular requires that a probe must scan over practically every part of the structure to be examined. A third reason why acoustic emission is a particularly attractive NDT tool stems from its value for the continuous in-service monitoring of structures. This is particularly important in offshore applications where other methods of inspection, which usually require divers, cannot be used in adverse weather conditions, while acoustic emission monitors will operate 24 hours each day, even in the worst storms when safety monitoring is particularly important. Strain gauges also have a '24 hour a day' capability but require to be placed very close to a suspected defect. The use of vibration monitoring of structures has also been developed in recent years. In this technique a frequency analysis of the natural vibration of a structure is recorded periodically. Any major change in the structure will be reflected in the vibration signature. Unfortunately, this method only indicates a really large fault — even to the extent that, in some cases, a whole member of a structure has to be removed before the fault is detected by this method.

Acoustic emission, of course, has disadvantages. The acoustic pulses have energies, in some cases, near to the lower level of detection of piezoelectric (PZ) transducers, which means that sophisticated and expensive electronic apparatus must then be used to detect faults. The main difficulty, however, is associated with the background noise which is frequently present during in-service tests of structures. Care has to be taken to reduce interference from noise generated in pumps, waves or moving machinery, and ways of overcoming this problem will be discussed later.

References

Drolliard T F 1979 *Acoustic Emission — A Bibliography* (NY: IFI/Plenum)
Frank F C 1949 *Proc. R. Soc.* A **62** 131
Frenkel J and Kontorova T 1938 *Phys. Z. SowjUn.* **13** 1
Kaiser J 1950 *PhD Thesis* Technische Hochschule Munich
Nichols R W(ed) 1976 *Acoustic Emission* (Barking, Essex: Applied Science)
Ying S P 1973 *CRC Crit. Rev. Solid St. Sci.* **4** 75

2 Acoustic Emission Techniques and Systems

2.1 Techniques

2.1.1 Introduction

The techniques of linear elastic fracture mechanics provide the design engineer with the information necessary to minimise the risk of unstable and dangerous fracture in structures, in terms of a maximum service stress coupled with a maximum allowable defect size. Even when the applied stresses are below those required to cause rapid crack growth and fracture, it is still possible for cracks to extend by a stable crack growth mechanism such as fatigue if the structure is subjected to fluctuating loads, or stress–corrosion if a corrosive environment is combined with stress. The practical conditions governing the initiation and growth of cracks are not always well understood, and thus it is advisable to inspect the structure periodically in order to detect any crack growth which may have occurred. For inspection purposes, conventional non-destructive techniques are available which can locate and size certain types of defects. However, these techniques are limited in applicability since large areas of a structure must be inspected. In addition they cannot identify the mechanism of cracking and are dependent upon the experience and judgement of the operator. Finally, some parts of structures are physically inaccessible for inspection purposes; clearly a remote method of locating, sizing and identifying the mode of propagation of defects in structures is desirable.

The atomic rearrangements which occur within a material during deformation and cracking produce elastic waves which travel through the material and can be detected at a surface by piezoelectric transducers. The transducer signals can thus be used to detect deformations or cracks in materials; this has practical application in the non-destructive testing of structures and components. It is possible, in principle, to obtain information about the nature and the severity of changes occurring at defects in structures under load; additionally, three or more transducers can be used to locate the position of the deformation within the structure or component. It must be stated, however, that the need for a skilled operator has not yet disappeared.

This is the basis upon which acoustic emission is used as a non-destructive testing technique. Figure 2.1 is a schematic illustration of this method. The equipment, which will be described in detail below, comprises the transducer, firmly bonded to the surface of the material under test, signal leads, whose design is important, connected to electronic signal conditioners, counters and recorders.

All the common materials used in structures (metallic alloys, glasses, polymers, ceramics and cements) exhibit acoustic emission. The transducer usually picks up a series of pulses of elastic energy rather than a

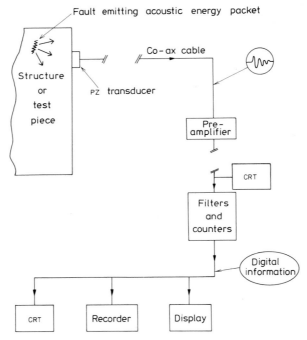

Figure 2.1 Diagram of a simple acoustic emission system.

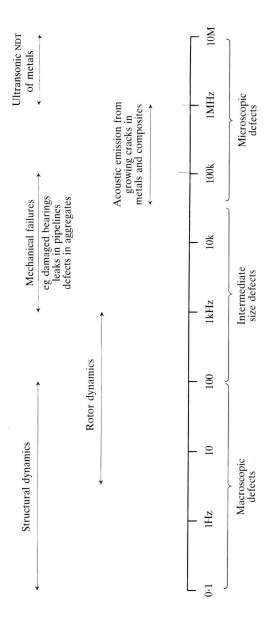

Figure 2.2 Spectrum of vibration and sound used for inspection and testing.

continuous wave. The energy and the frequency spectra of the elastic wave packets coming from the defect depend on the material, the nature of the deformation process, the local stresses at the defect, and often on the environment in which the material is placed. Figure 2.2 shows how the frequencies met in acoustic emission signals compare with those used in other common NDT techniques.

The most common deformation processes are due to:

(1) steady loads beyond the elastic limit of the material, causing either crack growth or fracture, depending on the material and on the localised stress;
(2) fluctuating loads, i.e. fatigue;
(3) fluctuating loads and a corrosive atmosphere or environment, i.e. stress–corrosion fatigue;
(4) crack growth in an adverse gaseous atmosphere, e.g. hydrogen-induced crack growth in metals.

Acoustic emission can be used to detect all of these processes.

Modern acoustic emission equipment can perform a variety of electronic functions on the transducer signals. It is usual to count the rate of production of pulses at the transducer, and modern equipment will estimate the energy contained in these pulses and in some cases go as far as giving a measure of frequency spectra. More sophisticated equipment can provide defect location by using an array of transducers, and attempts are often made to assess the severity of detected defects. Also, equipment at an experimental stage is being used to characterise the nature of the crack growth or deformation mechanism.

The present volume is mainly concerned with practical applications of the technique but it is worth mentioning that, as a large number of dislocations are usually involved in each burst of energy, theories have been mounted based on the assumption that the pulse can be regarded as a strain pulse in a linear elastic material. Despite work by Stone and Dingwall (1977) and Stephens and Pollock (1971), followed by recent theory by Hjelmroth (1978), little can be deduced about dislocation measurements and the generation of phonons from acoustic emission observations.

In general, materials and their environment have a profound effect on the rate of generation and the energy of the pulses. As a rule of thumb, brittle materials are 'noisier' owing to the higher energy per unit of crack growth compared with ductile materials, but this statement must be taken in conjunction with knowledge of the atmosphere in which the structure is placed. In general a corrosive or oxidising atmosphere can also produce noisy conditions. These statements will be amplified considerably in later chapters.

One of the limitations to the technique is the comparatively low energy in the pulses compared with other noise sources associated with practical structures. In-service testing of all structures must be carried out in such a way that background noise is not allowed to interfere with the measurements. The next chapter will discuss equipment needed to do this. My own experience has shown that any new application of this technique must be considered in relation to the 'noisiness' of the material for the structure, the environment and background noise sources. It is also necessary to take note of factors relating to acoustic transmission paths in the structure and the feasibility of leading signals back to the conditioning electronics. It is essential when starting an investigation to carry out small scale tests on the materials in question under conditions as close as possible to the service situation. This type of approach is well illustrated in the chapter on offshore applications.

When approaching a new possible application of acoustic emission the investigator must first of all find out the probable material failure mechanism, the environment in which the structure finds itself and the noise background. Tests in the laboratory can thus determine the energy of the acoustic pulses but care must be taken to simulate as far as possible the environment. The size and geometry of the structure will then determine if acoustic emission pulses from cracked areas could be picked up by transducers at the surface. If all is well and if noise can be filtered out then it is vital to carry out controlled experiments using test pieces coupled with calibration tests. Although laboratory tests are an essential in any acoustic emission investigation, great care must be taken in extrapolating their results to 'real life' tests; too much optimism — or pessimism — must not be allowed at the laboratory stage.

2.1.2 Information in acoustic emission signals

Emissions as received by the transducer contain information on:

(1) rate of emissions received;
(2) frequencies within the emitted pressure wave;
(3) amplitudes within the emitted pressure wave arriving at the transducer.

Additionally, energy parameters can be generated from the transducer signals.

The simplest way of characterising a pulse or series of pulses produced in an acoustic emission experiment is called 'ring-down' counting. Figure 2.3 shows the time–amplitude trace of a pair of typical signal bursts at the transducer. Counting the number of times per second the amplitude exceeds a preset voltage gives a simple number characteristic of the signal. An experienced operator can use this number to make observations concerning the severity of the rate of growth of a defect under study.

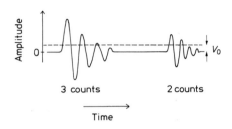

Figure 2.3   Typical acoustic emission signal bursts; V_0 is threshold for counter.

It will be noted that this simple approach relies on pulse counting, or in some cases, the measurement of an averaged signal amplitude. More sophisticated equipment, now generally preferred, adds energy analysis to simple counting because:

(1)   a 'ring-down' count is a function of signal frequency;
(2)   the count is only indirectly dependent upon amplitude because a large amplitude signal will often last longer than a low amplitude signal, i.e. the count is biased towards large amplitude pulses.

Energy analysis can mean any of the following:

(1)   the square of the initial pulse amplitude is measured for each burst;
(2)   the area under the envelope of the amplitude–time curve is measured for each burst;
(3)   the area under the actual amplitude–time curve is measured for each burst.

The electronic equipment normally used can then present a statistical analysis of the energy parameters measured on the individual bursts. Examples are given in the next chapter.

In practice most investigations have limited the frequency range received from emissions by a combination of resonance transducers coupled to bandpass filters. Some modern thinking tends towards the use of transducers with an essentially flat frequency response within the range 100 kHz to 1000 kHz coupled to bandpass filters around 100 Hz wide, as the use of frequencies over too wide a band is susceptible to excessive background noise, limiting sensitivity and resolution of the technique. The use of broad-band systems is also receiving attention and the use of such systems is described later in the chapter.

At this stage it must be noted that when a stress wave propagates through a specimen or a structure it is subject to modification from discontinuities, such as welds, or from reflections at free surfaces (particularly when there are sharp changes in the specimen geometry), or by multiple reflections. As a result the form of the acoustic emission as recorded by the transducer, which has to be placed at a surface, may well differ from the form of emissions actually generated by the deformation or cracking process within the specimen. The amplitude distribution of received emissions is not necessarily the same as that starting from the stress point, as the recorded emissions are highly sensitive to the geometry of the structure, the location of the transducer and the characteristics and dimensions of the particular detector used. Despite these facts it is normally assumed that the proportionality between emission signals will be the same for each cracking mechanism. Some authors have attempted to examine the way in which the perceived signals relate to true emission pressure pulses.

2.1.3 Defect location

It is evident that by using three or more transducers, the position of a source of acoustic emission within a structure can be determined by measuring the time of arrival of bursts at the transducer. While the next section contains a full description of two defect location systems, it is useful to briefly indicate the principle of operation. Figure 2.4 shows the 'centred triangle' array of four transducers often used. A large structure will have a number of such arrays spread over its surface. Usually the triangle or star configuration is used when locating on a flat

surface, and a diamond pattern on a cylindrical surface. These arrays
can be randomly placed about a structure as each has its own coordinates
and monitoring area, totally independent of the others.

The arrival time of acoustic signals at each transducer is measured
with a resolution in the region of 0·1 μs. The arrival time at each
transducer is compared with the others and so the position of the source

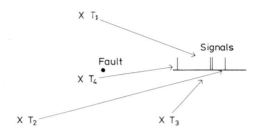

Figure 2.4   Centred triangle array.

can be calculated once the velocity of sound in the material is known.
This velocity is usually determined experimentally by arranging for a
sharp mechanical pulse to be applied at a number of new positions on
the structure during calibration tests.

2.2   Systems

2.2.1   Introduction

This section is concerned with systems aspects of acoustic emission. It
seeks to describe the equipment used in the many varied applications
of this technique, ranging from laboratory studies in a tensile test
machine through to on-site applications in extremely adverse environments such as, for example, the use of acoustic emission to monitor
offshore structures.

The system design depends on three main factors: the physical location
of the equipment, the type of investigation (i.e. laboratory studies,
structural integrity monitoring) and on the way in which the final information is to be used. There are, broadly speaking, two types of system,
depending on whether defect location is required or not. In laboratory

studies where the position of the fault is well known (i.e. in a tensile test specimen) a simple, non-locating system may be used. However, the proof testing of a structure or pressure vessel requires a large computer-based installation with location capability. In the latter case it is also necessary to provide the electronic equipment needed to assess the severity or nature of a fault. There are also instances where source location coupled with simple assessment of severity is more important than the more sophisticated analysis needed to characterise a fault. This probably represents the most important area of industrial application.

The basic, non-location acoustic emission detection system was shown in figure 2.1. This type of system will now be discussed in some detail. Location systems will then form much of the latter part of this chapter. A good example of typical simple laboratory equipment has also been described by Speich and Schwoelbe (1975) in STP 571 who used their apparatus to carry out transformation studies in steels. It is similar to simple commercial equipment (for examples of which see the trade literature of the Dunegan Endevco Co. or the Trodyne Co. or others). A lead zirconate PZ transducer resonating at, say, 150 kHz is used with a bandwidth of 200 Hz followed by a wide-band amplifier. The form of the signal from the PZ transducer is as shown in figure 2.3 and can be approximated to a damped sine wave. A digital–analog converter is used before passing signals on to a plotter (for some experiments amplified signals are passed through an envelope processor which is used to combine multiple reflected signals into one pulse). The method of signal conditioning and counting in widespread use with this equipment is ring-down counting. Many investigators have adopted slightly different procedures for ring-down counting but the equipment described in the article by Speich and Schwoelbe represents a basic and widely used system.

2.2.2 Energy analysis

As discussed in section 2.1.2, ring-down counting is now less common, and is being replaced by energy analysis of acoustic emission pulses. The measurement of the energy in a signal by means of electronic processing is, in principle, simple. The signal voltage is first squared, and then the area under the curve of voltage squared against time is measured. This area is proportional to the signal energy with the constants of proportionality being the amplifier gain and input impedance.

Figure 2.5 (*a*) shows a block diagram of a circuit to measure energy. The squaring circuit is a commercially available integrated circuit multiplier. The area-measuring circuit is shown in figure 2.5 (*b*) and is a type of voltage-to-frequency converter. This circuit is described, along with design criteria, in some detail by Beattie (1976). In this voltage-to-frequency converter, the capacitor is charged with a voltage V_0 which is proportional to the squared signal voltage. When the voltage across it reaches a test voltage, V, the capacitor closes the switch, discharging the capacitor, and also sends one pulse to the counter. If the capacitor charges linearly, each pulse will represent an increment of area under the voltage squared against time curve. The amount of energy represented by each pulse will depend upon the circuit parameters and amplifier gain.

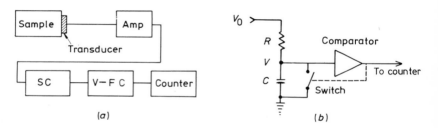

Figure 2.5 'Energy' detection system: SC, squaring circuit; V–FC, voltage-to-frequency converter.

If the capacitor is discharged by a negative pulse of known characteristics instead of a switch (in this case a FET), the device becomes a charge pump. The characteristics which must be considered in choosing an area-measuring circuit are linearity, dynamic range and maximum speed. The maximum pulse repetition rate of the circuit is important as it determines the resolution of the device. For example, if near the end of a test, large emission bursts occurred at an average rate of 50 kHz and it was desired to have resolution such that a smaller burst with 100 times less energy would get one count, a maximum repetition rate of 5 MHz would be needed. The wide range of acoustic emission signals encountered makes a minimum repetition rate of at least 1 MHz desirable. To date, the best dynamic range achieved in this type of circuit is about 25 dB. A circuit of this type is commercially available.

The Dunegan Endevco Model 602 is also equipped with energy processing, which takes the maximum transducer signal, squares it, filters it and produces a series of digital pulses stored in a totaliser giving a reasonable approximation to the true energy of the transducer signal.

A comprehensive description of equipment based on a different approach to energy measurement (described below) has been given by Bartle (1975) who gives a credit to the Admiralty Materials Laboratory (AML) which played a major role in advancing the development of detection and amplification systems under the direction of Birchon (1976). In the AML equipment, initial amplitudes of signals are changed to squared amplitudes electronically and then summed. The PZ transducer chip is also electrically connected to a pulse transformer to match the chip's output to that of the special low-impedance cables used. Other workers have also adopted energy assessment including Mirabile and Palombis (1973) at ATEL. Work concerning energy assessment has also been carried out by Watkins (1972) and Beattie (1976).

The complete AML–Welding Institute equipment comprises a two-part assessment system based on detection and amplification units. The transducer consists of a standard PZT5A piezoelectric chip, mounted in a heavy brass housing to provide adequate shielding from airborne interference. The housing also contains a toroid, matching the impedance of the chip to that of the 'super-screened' cable (see Birchon 1976) used to connect the transducer to the preamplifier. This cable is a vital feature of the equipment. 0·5 MHz resonance frequency chips have been used in the main, but occasionally a wider range of transducer frequencies is employed to reduce difficulties in detecting emission from some sources. The preamplifier has a wide bandwidth (150 kHz–1 MHz), a low inherent noise characteristic (0·2mV/Hz) and a fixed gain of 20 dB. To obtain the full benefit of these capabilities the preamplifier is also enclosed in the brass housing. The main amplifier cascade is constructed to a low-noise design. Its filters are switchable to 120, 175, 263, 394, 590 and 885 kHz centre frequencies, ±20% bandwidth. At all but the maximum frequency range, this allows the ambient electrical noise level to be maintained equivalent to about 2 μV zero-to-peak at the preamplifier input, but at the maximum frequency this value is sometimes about 4 μV.

The assessment system gives values for emission energy as the sum of the squares of the peak emission amplitudes (A^2) at a selected frequency, as well as the corresponding amplitude distribution. A^2 values are displayed, referred to 1 μV at the preamplifier input as unity. The minimum signal recorded is variable upwards from 3 μV (at the preamplifier input) in 10 dB steps. Normally one channel is set for maximum sensitivity and the other to accept only larger emissions; this is either to give an additional breakdown of the relative significance of larger and smaller emissions, or to allow emission to be detected above and below ambient mechanical noise levels, i.e. excluding and including mechanical noise.

In a notation used by Bartle (1977), where appropriate the dynamic range monitored is shown as the limits on the summation sign, the unit value used as a subscript to A^2 and the frequency monitored as a suffix in brackets. Thus

$$\sum_{3\,\mu V}^{3\,V} A^2_{1\,\mu V}\,(2 \cdot 0\,\text{MHz})$$

means that the values refer to a dynamic range of 3 μV–3V to a 1 μV unity, and the centre frequency of the detection system was 2·0 MHz. The amplitude sorting system has ten channels of 10 dB separation, the most sensitive counting emissions within the amplitude range 3–10 μV at the preamplifier input. The highest channel has an open upper end. The number and separation of the channels was chosen as a compromise between obtaining adequate discrimination and obtaining too many sets of data to be readily handled. The relatively low minimum level of the highest channel (100 mV) reflects the sensitivity of the transducer (when emission is readily detectable), since emissions of greater amplitude than this occur relatively rarely and indicate that major damage has taken place whatever their true size. All the data displayed, including running time and an event marker number, are printed out either at preselected time intervals (3, 10, 100 or 1000 s) or on demand. The triggering circuits are designed to allow optimum operation in the more demanding circumstances of variable emission amplitude and high frequency of occurrence of emissions. In particular, if two events occur such that one is being assessed when the second occurs, then the larger (and more significant) is assessed if it is not possible to assess both. Additionally, an arbitrary count rate is recorded when the emission rate is sufficiently high that it prevents a trigger level from resetting. It should also be noted that this equipment assumes a constant decay time for each emission signal and also incorporates a dead time into the circuits.

Bartle has examined the relevance of A^2 as compared against the A and A^3 methods of assessing a series of acoustic emission signals. In figure 2.6 results from a stress–corrosion test (Bartle 1974) on an aluminium alloy are plotted so that the measured projected area of cracks found when the specimen is broken open is plotted against A and A^2 (for clarity the results for A^3 are not reproduced here). A^2 allows a curve to be drawn which is within all uncertainty bands where crack areas have been positively identified. It can be seen from the original paper that the A and A^3 plots show some areas of very poor correlation. Bartle maintains that the correlation between assessed values and crack areas

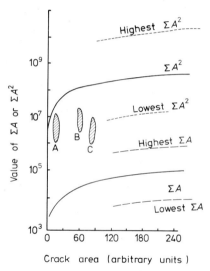

Figure 2.6 Comparison of $\Sigma\,A$ and $\Sigma\,A^2$ assessment of crack growth. Values of $\Sigma\,A$ and $\Sigma\,A^2$ have been plotted against crack area (measured by cutting open specimens). $\Sigma\,A$ shows poor correlation with area for areas between 0 and 100 units. $\Sigma\,A^2$ has no such 'bad' regions as A, B and C.

is as good as can be expected with the A^2 assessment approach. He also maintains that the A approach is invalid since it overweights (relatively) small emissions, while the results indicate (rather than prove) that A^3 assessment is invalid because it underweights (relatively) small emissions.

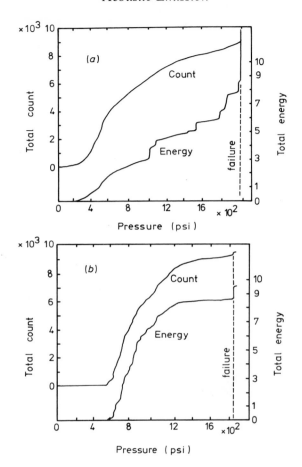

Figure 2.7   Energy and total count curves from two epoxy bottle tests.

Beattie (1976) has carried out a detailed, experimental comparison of ring-down and energy counting and gives an example of the usefulness of energy analysis from burst tests of a PRD†–epoxy filament-wound bottle. The data were recorded on a wide-band (0–50 kHz) magnetic tape recorder and analysed after the test. Figure 2.7 (*a*) shows the total energy and total count curves for this test. As can be seen, seven events

† A form of glass fibre.

before failure contained 50% of the energy in this test while these events can barely be seen on the total count curve. To see why there is this difference between the energy and count curves, one must examine the waveforms of the individual emissions. These signals had a frequency content centred around 200 kHz and were about 76 µs long. Thus each event recorded about 15 counts. During the course of the test about 6000 of these events occurred with an average repetition rate of one every 20 ms. Two of the seven high-energy events, which had a pre-dominate frequency of 35 kHz, were about 2 ms in length and thus recorded around 70 counts each. Comparing the two types of events we see that the high-energy events had a count approximately five times larger than the low-energy events while the energy content was over 800 times greater.

Examination of the bottle after the test showed several areas, in addition to the region of failure, where lines of broken filaments indicated the occurrence of propagating cracks. This strongly suggests that the small events were produced by either matrix cracking or the breaking of individual strands in the filaments, while the high-energy events were produced by propagating cracks. If this supposition is correct then energy analysis in this experiment easily differentiated between acoustic emissions associated with events which had little, if any, effect on the strength of the bottle and acoustic emissions associated with structural failure.

Figure 2.7 (*b*) shows energy and count curves from a similar PRD–epoxy bottle. This bottle differed from the first one only in that a different epoxy system was used for the matrix. The two curves show no essential difference between the total counts. Examination of the waveforms showed that all were similar. Finally, this bottle showed no indication of damage occurring in any region on the bottle except the region of failure. In the region of failure, damage characteristics were similar to those exhibited by brittle fracture, unlike the first bottle where the damage appeared to be tears in several layers of the windings.

The experimental results discussed in this section show that energy analysis is an effective method of differentiating between acoustic emission signals which have different frequency and damping characteristics. In the case of the first PRD–epoxy bottle, energy analysis easily differentiated between emission signals arising from what appear to be structurally harmless and structurally dangerous processes. Thus, energy analysis is a very useful tool in cases where emission signals with several different sets of characteristics appear in the same test. These differing signals can be produced by different materials in the same structure. A prime example of this is welds, where the difference in material or heat treatment in the weld may cause the generation of emission of a character different from those generated by the base metal.

A comparison between energy analysis and ring-down count analysis has also been conducted by Harris and Bell (1974, 1977). They found little difference between energy analysis and ring-down count analysis for crack growth in 7075 T6 aluminium, 4341 steel and 3·9% carbon nodular cast iron. Only in unflawed tensile tests of the aluminium did energy analysis appear slightly better than ring-down count analysis. These studies were quite thorough but subject to one criticism. In almost all cases, the electronic bandpass was limited to 0·1–0·3 MHz. This tends to eliminate one of the prime advantages of energy analysis, that of being frequency independent.

Finally, it must be pointed out that in cases where there is only one generating mechanism and/or the emission signals all have similar wave-forms, energy analysis appears to have no advantage over ring-down count analysis. In these cases, the added complexity and limited dynamic range of energy analysis make ring-down count analysis the preferred method.

2.2.3 Transducers

Transducer characteristics are an important feature of the specification of acoustic emission equipment. Reference will again be made to trans-

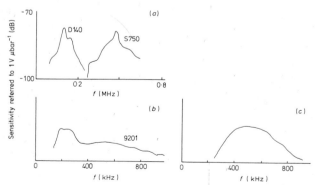

Figure 2.8 Frequency responses of acoustic emission transducers: (*a*) typical, (*b*) wide-band, (*c*) noisy environment.

ducers available from Dunegan Endevco and from AML. The frequency response of two general purpose transducers are shown in figure 2.8, both having a peak sensitivity of -75 dB referred to 1 V μbar^{-1}. The frequency response of a wide-band transducer is also shown in the figure with a further curve showing the response of transducers specially designed for the measurement of acoustic emission in the presence of a high level of mechanical noise. The sharp decay in sensitivity at frequencies below 150 kHz has made them insensitive to most mechanical noises. A high sensitivity transducer has also been produced having a maximum sensitivity of -65 dB referred to 1 V μbar^{-1}.

2.2.4 Interference signals

The workers at the Admiralty Materials Laboratory have also studied the frequency spectra of signals competing with stress wave emissions. This work has been reported by Birchon (1976) and figure 2.9 is taken

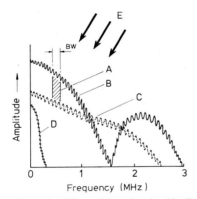

Figure 2.9 Schematic of the problems involved in discriminating the desired signal from the background: BW, bandwidth; A, available signal; B, transducer response from true signal; C, typical flow noise interference; D, typical mechanical noise interference; E, electromagnetic interference.

from his paper. It shows the frequency band of a normal acoustic emission system compared with the usual transducer response, electromagnetic interference, mechanical noise interference, and interference from flowing water or gases. Mechanical noise, such as that from testing machines decreases rapidly in amplitude above 100 Hz while flow and cavitation noise can produce frequencies as high as 2 MHz or more. So it is possible to design a testing system which, of the types of interference shown in figure 2.9, will not be sensitive to mechanical or flow noise. This, although not always easy, can be achieved by ensuring that transducers are fitted in such a way that interfering noise is attenuated by welds or by layers of plastic, or other means. It is also sometimes possible to ensure that mechanical vibrations are eliminated when the measurements are being taken. Flow noise causes more problems but there are many cases where pumping machinery can be turned off while a test is being made, eliminating many interference sources at the same time. Electrical interference is not easy to eliminate; in one series of tests (with which the author has been associated) at a place well away from power lines and radio transmitters, high-frequency electromagnetic pick-up proved to be a serious source of difficulty.

Birchon has discussed the use of special low-interference cables devised by Wilson and Fowler (1973). In the case of normal coaxial cable the switching of a mains voltage near the cable can induce 'scream' currents of up to 100 mA at the resonant frequency of the cable. Their observations led them to design multishielded metal and couple screened cables with a surface transfer impedance some 10 000 times better than normal coaxial cable. The use of this is an essential part of the AML defect location system and of the equipment produced by the Welding Institute. The results of Birchon illustrated graphically how important it is to give attention to cables. Birchon claims that even a 100 m length of cable between transducer and preamplifier can be used with impunity and he claims significant improvement in this practice compared with the normal use of a preamplifier close to the transducer. A discrimination ability of the order of 1·5 V at the transducer is claimed, compared with the typical 10 V level met in a reasonably well screened laboratory in the absence of special cable.

Further work on background noise, concentrating on reactor applications, has been carried out by Hutton (1971) and by Ying *et al* (1974). Hutton's work concentrated on the measurement of the frequency spectra of hydraulic noise as detected on a test hydraulic loop consisting of 66 mm steel tubing containing water flowing up to 9 m s^{-1} at a temperature of 260 °C. Cavitation was induced by a gate valve 3 m upstream of the point where an artificial acoustic emission generator was attached. The emission generator comprised an aluminium bar subjected to mechanical stress, in contact with hydraulic loop piping. The frequency spectrum of the background noise was peaked at around 100 kHz with a small secondary peak at around 400 kHz. The use of the 300 kHz high-pass filter reduced the background noise to a low level apart from a small peak just below 400 kHz. Hutton concluded that the detection of acoustic emission above 1·5 MHz appears to be unaffected by hydraulic noise or cavitation. He maintained that all of the emission from steel consists of sharp rise transients producing frequencies in the megahertz region. He further recommended the use of an edge-mounted shear-wave sensor, both for the resolution, compared with background signals, and the accuracy of source location. He found that a surface-mounted longitudinal-wave sensor was not suitable for use when cavitation noise is present.

Ying and his co-workers (1974) studied the noise structure of sound associated with a boiling water reactor (BWR) and comparing this with the available noise data from a pressurised water reactor (PWR). At

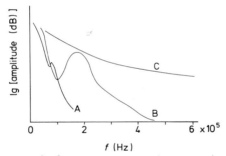

Figure 2.10 PWR noise frequency spectrum: A, pump noise; B, 'total' PWR noise; C, frequency analysis of acoustic emission signal envelope.

frequencies near 500 kHz the hydraulic turbulence and cavitation noises of the PWR were about 0 dB for the selected normalisation and the noises of the BWR varied from 0 to 15 dB for the various locations that were monitored. On the same scale, acoustic emission signals lie in the range 0–20 dB so that the PWR background noise should not completely mask these signals but the situation in a BWR would be more difficult. As in the work of Hutton, a low-frequency peak is associated with small peaks at around 300 kHz. The acoustic background noise on a vessel flange and on one loop of a PWR are shown in figure 2.10. In general all spectra show high peaks below 100 kHz due to high mechanical and hydraulic noise with some noise above 100 kHz due to turbulent flow and cavitation noises.

In some cases acoustic emission has to be detected against a background of machine fatigue noise. In a study of this situation Horak and Wehreter (private communication) designed and tested an acoustic emission system which integrated known noise discrimination techniques with computerised source location equipment. These discrimination techniques comprise master–slave, coincidence and rise time detection. The application of this integrated system to progressively more noisy aircraft components, showed that 95–99% of all noise external to the transducer array was rejected by the hardware prior to computer processing. Crack initiation and propagation were detected in high-strength steel elements undergoing severe fatigue testing. The detection and location of crack initiation and propagation would have been virtually impossible in this environment without noise discrimination.

Hydraulic components subject to fatigue cracking can be difficult to inspect using periodic proof testing if sources of 'apparent' acoustic emission are present. A paper by Wright and Washburn (1976) defines a systematic approach for noise identification and suppression used in the development of a successful inspection technique on an actual hydraulic component. Techniques were developed for noise suppression using variable system gain and frequency range, and a test method for quieting moveable seals was presented. The result was a technique appropriately insensitive to these noise levels, yet with a significant gain margin for the detection of fatigue-crack related emissions.

2.2.5 More sophisticated systems for eliminating noise effects

So far we have in the main dealt with the use of frequency discrimination as a method of avoiding unwanted background noise which can mar or distort acoustic emission signals. The severe conditions met in aerospace and offshore applications have led to the development of more sophisticated systems for the elimination of the effects of noise. In aircraft, moving rivets cause severe noise problems and workers at the Royal Aircraft Establishment (RAE) at Farnborough have developed a most useful guard ring method. This and the other approaches, using velocity validation and other location techniques, will now be discussed.

(*a*) *Guard ring method.* The RAE guard ring equipment (P. Dingwall, private communication) comprises a four-level amplitude sorter designed at the RAE in conjunction with Surrey University. This system uses half-wave rectification and is designed such that the most sensitive trigger level in the sorter is fired at a transducer output of 68 μV peak-to-peak. The other three levels are fired by signals of 20 dB, 40 dB and 60 dB above the guard amplitude (the bandwidth is 70 kHz to 1 MHz; 3 dB points). Two channels are used in the amplitude sorter, one being the main signal channel. The second has three transducers connected to it, arranged in a ring around the signal channel. Logic in the equipment is such that a stress wave reaching any one of the three guard transducers before the the signal transducer, inhibits the output of the signal channel until the stress wave has had time to leave the signal transducer. Thus the field of surveillance of the signal transducer is confined to a region around it with a radius of half the distance between guard and signal transducers. During one 10 minute test run 70 000 counts were obtained on the signal channel while the guard channel registered 5·2 million counts. Very high level signals of 70 mV peak-to-peak or greater were obtained and a crack was detected. Results from this equipment are reported later but it can be seen how an extremely noisy structure can be tested using this approach.

(b) Velocity and parametric validation methods. In my opinion the most effective method of determining emissions in the presence of considerable background noise uses velocity validation. This has been used successfully for the detection of emissions in offshore situations where the level of background noise is very severe (C H J Webborn, private communication). The Dunegan Endevco 1032 system or the EMI system, for example, determine the difference in arrival time of pulses from emissions, either genuine or spurious. These values are normally called Δt values. This system in common with other sophisticated modern acoustic emission equipment, in addition to measuring Δt values, also logs the amplitude, rise time, duration and ring-down counts of all signals received (with the exception of those during a short dead time). This information can be displayed on a CRT in real time in a variety of forms (e.g. number of events of amplitude A plotted against amplitude or the logarithm of the number of events against amplitude). The signal arrival time is measured with a resolution of 0·1 μs and is used to plot the position of origin of all emissions. Also included is the facility for gating the analogue signals via a parametric input, such as a load,

measured either on fatigue equipment or from wave data or from any other useful parameter. This permits data acceptance only during those parts of any cycles of the structure or equipment under test when background interference is minimal. Alternatively the same facility can be used to guard the sensors from remote sources of background mechanical noise in a similar way to the RAE equipment described above.

The method of filtering out unwanted signals in this equipment operates on the principle that emissions are only counted if: (1) their source lies within the area determined by linear location between two sensors; and (2) the emissions occur when the background noise as determined by the voltage-controlled gate (the parametric gate) is at a low level.

Extensive tests both in the offshore situation and also in large-scale fatigue tests at the National Engineering Laboratory have shown that this method of combined parametric and velocity (or spatial) filtering is most effective in determining which emissions: (1) are of interest in diagnosing the state of the structure under test; or (2) are derived from extraneous sources, such as wave motion or mechanical noise from rig machinery.

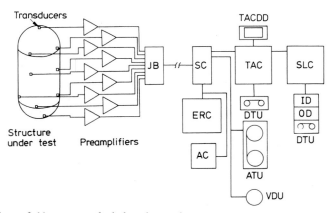

Figure 2.11 ACOUST fault location and assessment system: JB, junction box; SC, signal conditioning; ERC, energy release circuit; AC, audio circuit; TAC, time analysis computer; TACDD, time analysis computer data display; SLC, source location computer; ID, input device; OD, output device; DTU, digital tape unit; ATU, analog tape unit; VDU, visual display unit.

2.2.6   Field testing equipment

Many authors have described equipment which is purpose-built for field testing. Reference can be made to two papers, one by McElroy (1975) on the examination of gas piping buried in the ground and the other by Parry (1975) in which the portable ACOUST system of the Exxon Corporation is used for commercial acoustic emission testing. Figure 2.11 illustrates the pipeline system which is very similar to that used by Exxon. This uses energy discrimination as an early warning system for serious failure. The equipment also incorporates a time analysis computer whose operation will be described in more detail below.

2.2.7   Defect location systems

The major development in acoustic emission systems in the last few years has been in systems to locate as well as to verify and estimate the severity of faults. Such systems will be discussed initially by reference to two examples, both designed for the field testing of steel pressure vessels. The first equipment to be discussed was developed at the Berkeley Nuclear Laboratories of the British Central Electricity Generating Board. A particularly full account of the development and operation of this equipment has been given by Sinclair (1977). This system is referred to by the acronym ACEMAN (*acoustic em*ission *an*alysis). The CEGB required the system to operate in environments having a considerable background noise. They also wanted to have a real time display including mapping and amplitude analysis of emission signals. The system is designed to avoid loss of data due to 'lock out' when two or more signals arrive simultaneously at the transducers. Sinclair and others also claim that the system can be used with any sensible array of transducers.

ACEMAN uses up to 12 sensors operating on a 165 kHz band centre; the electronic system prints out the amplitude of each emission plus the time of each emission as received by the transducer array. Timing can be either at signal peak (used for simple structures) or at a leading edge when more complex arrays are being used. Source location is carried out by noting that sound from defects travels at approximately 3 km s^{-1}(the speed of Rayleigh waves). Consequently, for an array of three transducers one pair of arrival times will define the difference in distance from the emission source of the transducers. These therefore define the foci of a pair of hyperbolas and since two pairs of transducers are used the source is located from the intersection of the hyperbolas. The ACEMAN system uses the exact solution method for the computation described by Tobias (1976). In practice the transducers are placed at each vertex of an equilateral triangle with a fourth transducer positioned at the centre of the triangle.

To minimise data-handling problems and to enable real time readout, a fault test is preceded by the computation of a table from which source location can be inferred by reference to the computer store. Normally the table has an array of 64×64 elements and consequently the resolution of the system is 1/64th of the side of the equilateral triangle. Acoustic events are stored on a magnetic disc which is capable of handling 1.5×10^5 potential events. Amplitude spectra for selected events can be obtained.

When a survey is being undertaken it is frequently necessary to use 3 mm acoustic waveguides in order to couple the structure to the transducers. These are designed to maximise pick-up from surface Lamb waves with stray components perpendicular to the surface. Sinclair and others also discuss the question of a maximum source–transducer distance and conclude that this should not be greater than 6 m.

Background signals from mechanical or hydraulic noise are also received by the acoustic sensors. Fortunately, emissions associated with the impulsive events of crack growth have characteristically sharp rise times. The rise time as observed at a sensor is lengthened according to the distance from the source, the distribution of source energy into the propagating wave modes, the velocity and dispersion of individual modes within the receiver bandwidth, and mode conversions and reflections at any geometrical discontinuities in the sound path. Nevertheless in many practical circumstances the signal from defect growth retains a sharp rise time. This provides useful discrimination against the more slowly rising signals associated with extraneous plant noise.

As an additional guard against spurious emission sources, four sensors are normally used in each sub-array of the ACEMAN system. Three sensors are required for the location calculation while the fourth validates the emission event. The signal arrival time measured at this fourth sensor is compared with that predicted for the calculated source location. In this way chance coincidences between signals from different sources, the effects of sound propagation by multiple paths in closed structures, and source location ambiguities are effectively eliminated.

Along with event detection and timing, the peak amplitude of the emission event is measured, together with the instantaneous value of a parameter, such as pressure, appropriate for the test. These data permit detailed evaluation of identified emission sources and correlation of activity with parameters relevant to fracture mechanics. The main use of ACEMAN in the field has been in connection with nuclear pressure vessels, and it has been observed that the simple relationship $M = bV$ between the total count and yielded volume obtains, with b as low as 2×10^{-3} events/cm^3.

The Exxon Corporation operate a portable location and analysis system known as ACOUST. This has also been described by Parry (1975) but with fewer details than that of the CEGB system described above. Fault diagnosis is carried out with reference to accumulative energy count and to time analysis of received signals. No details are given by Parry of the method of computation required to locate the source but it is similar to that adopted by Sinclair and others. The output from this system is given in the form of a digital map where the numbers nought to nine relate to the energy readings and therefore to the severity of defects.

In all systems the computation of source location is often done by using a pulser in order to provide experimentally determined source location data.

Location systems are also made and sold commercially by the Welding Institute of England, EMI, and by three American companies, Dunegan Endevco, Acoustic Emission Technology Corporation and Trodyne. The South West Research Institute also build location equipment according to customer specifications. The Westinghouse Corporation have recently developed their own system for use in testing nuclear installations; it is described by Gopal in STP 571.

One of the features of modern location systems is the sophisticated software normally supplied, typified by the Dunegan Endevco 1032 range. The system is based on a four transducer configuration in the form of a centred equilateral triangle. Multiple arrays can be positioned randomly on a structure, as each array can have its own size, monitoring and wave speed calibration programs. Real time VDU presentation of sources is available with an indication of their severity. Operation is normally started with a dual histogram indicating the number of events per array and the counts per fault. The events per array histograms are always retained in the computer memory. The system uses data from a pulser calibration exercise to give defect location. A teleprinter makes available data such as time intervals, source coordinates and the number of counts for an event. This equipment now includes a floppy disc facility to save time in data acquisition. Additionally, the CRT display can include:

(1) A planar display showing location of events relative to transducer locations for any array.
(2) A linear locational display providing a histogram of number of events on the vertical axis against linear location on the horizontal axis.
(3) A histogram of activity on all arrays displayed simultaneously with location information. The histogram data includes *both* the number of events per array and the number of counts per array.
(4) A disc marker indicates events recorded on disc and events displayed.

Experience with the use of acoustic emission location systems in the presence of background noise has shown that care has to be taken over the positioning of transducers. If the source of noise is readily located and is restricted this task is simple. However, there are cases where background noise enters the structure over the whole of its surface near to the weld or suspected fault region. In this case it is important to try to locate transducers asymmetrically with respect to any suspected danger area, otherwise the spurious fault will be located at the weld.

2.2.8 Calibration

It is important to have access to a source simulator for many reasons, probably the most important being that of calibrating and checking the operation of source location systems. Many workers have devised simple methods of providing a simulated source which can be easily used in the field and which give signals comparable in frequency and amplitude to those from real emissions. Birchon and his co-workers pioneered the use of a simple aluminium block, one surface cut into perpendicularly and subjected to mechanical strain by means of a screw, from which, when salt solution is put into a crack, stress–corrosion signals are produced very similar in nature to those from natural cracking phenomena.

An ingenious alternative source has been described by Nielsen (1970) which uses a standard propelling drawing pencil of a commercial type. A lead of 0·5 mm thickness is brought forward by pressing a button at the far end of the pencil so that the lead protrudes by a standard distance, say 3 mm. The pencil is guided obliquely towards the surface of the steel structure until the guide ring part of the pencil body is resting on the surface. The pencil is then turned round the point of contact, towards the vertical, breaking the lead. Then the pencil lead is levelled in preparation for the next breaking action. It is convenient to standardise six steps in the protrusion of the pencil lead before the breaking process. It has been shown by Nielsen that this method produces a remarkably consistent signal. The quality of the lead had no significant influence on the amplitude of the generated signal, in this investigation. Changing the protruding lengths of lead, and the angle between the lead and the steel surface has little effect on the signal amplitude received. These signals were shown to be of the order of 60 dB higher than RMS noise in the system used by Nielsen. In the case of a steel material Nielsen has made detailed numerical results available in an internal paper from the Risø National Laboratory of Denmark.

A further important source of emission signals, which has the capability of being a secondary standard, has been described by Green (1978). The inability to determine the absolute sensitivity of acoustic emission transducing systems has been tackled by a simple low-cost method for the measurement of the absolute signal-to-noise ratio, namely the use of white noise energy as produced by a stream of helium gas hitting a specimen surface. This method allows the direct comparison of system sensitivity in the range 0–1 MHz regardless of the type of signal handling equipment used. This method was first suggested by MacBride (1976) but Green's paper gives more detailed results.

2.2.9 Very broad-band detection systems

The detection systems described so far in this chapter operate in what is the natural frequency range to optimise signal amplitude and minimise background noise, namely in the 15–300 kHz region. Several authors have pointed out that in order to utilise information about the shape of the emission pulses, as opposed to their amplitudes, it is necessary to use a very broad-band transducer going up to the megahertz region. Notable amongst these is Curtis (1974) who in a useful review paper has

described a system used at AERE Harwell capable of flat response up
to 10 MHz. This transducer is based not on the PZ effect but upon the
variation in capacitance of a capacitor when the distance between the
plates is varied. One plate of a small plastic-filled capacitor is made to
adhere to the specimen while the other is attached to an inertial mass.
Curtis has demonstrated that this device is linear up to 10 MHz and has
described a few simple illustrative examples. The use of a similar air-
filled transducer has been developed by his co-workers, Scuby and
Wadley.

Elementary physics shows that the output signal E of the charge
amplifier of a probe of this nature is given by

$$E = -k\,(\varepsilon A V/\xi)\delta\xi, \qquad (2.1)$$

where ε is the dielectric constant of the gap between the plates, A the
area of the plates, V the plate voltage, and ξ the plate separation. Using
thin metallised plastic sheeting (such as 25 μm thick Melinex) provides
a very good transducer which can adhere closely to a specimen of any
shape. An inertial mass is provided to some extent by the pick-up
electrodes. Substitution of relevant numbers into equation (2.1) shows
that a capacitance device has a sensitivity to surface displacement some
170 times less than a quartz chip at its resonant frequency. However,
a foil used shortly after application of a polarising voltage is only 50
times less sensitive than a PZT5A ceramic transducer. Examination of
this surprising result showed that this (and other phenomena such as an
aging effect) could be attributed to electret behaviour. This convenient
effect makes the use of a metallised Melinex foil quite practicable in a
laboratory situation. The foil displacement is determined by a charge-
following device, frequently used in nuclear applications e.g. semicon-
ductor particle detectors. The sensitivity of this transducer as a function
of frequency has been examined by Curtis using a nickel rod vibrating
against the transducer, energised by a magnetostrictive coil. This method
of checking the system cannot be used above 2 MHz; up to 5 MHz or
thereabouts the transducer was placed either side of an aluminium block.
As the attenuation of sound in polycrystalline aluminium is 6 dB per
octave below 5 MHz, the recorded decrease of 7 dB per octave suggests
that the transducer response is linear from 200 kHz to 5 MHz. Specimen
replacement causes no damage for a few tests. Its use outside the

laboratory is greatly limited by electrical interference and Curtis had to take elaborate precautions to overcome background noise problems.

References

Bartle P M 1974 *Stress Wave Emission Monitoring, Inst. Mech. Engineers Conf. Publicn.* 8
——1975 *Welding Institute Res. Bull.*, Sept. 1975 p255
——1977 Welding Institute, ECSC Contract
Beattie A G 1976 *Mater. Eval.* **34** 73
Birchon D 1976 *Br. J. Non-destr. Test.* **17** 66
Curtis G 1974 *Non-destr. Test. Int.* **7** 82
Green G A 1978 *Non-destr. Test. Int.* **11** 69
Harris D O and Bell R L 1974 *Dunegan Endevco Report* DE–74–3A
——1977 *Expl Mech.* **17** 347–53
Hjelmroth H E 1978 *Report of ECSC Project* 7210/GA/9/901
Hutton P H 1971 *Battelle, Northwest Report* BNWL 1597
MacBride S 1976 *Can. J. Phys.* **54** (17)
McElroy J W 1975 *ASTM, STP*571 p59
Mirabile M and Palombis E 1973 *Proc. 3rd Int. Conf. Fracture Mechanics, Munich* 1973 (Berlin: VDEn)
Nielsen A 1970 *RISØ Report No. M–1970 RISØ Danish Atomic Energy Research Establishment, Rosskilde, Denmark*
Parry D L 1975 *ASTM, STP*571 p150
Sinclair A C E 1977 *CEGB Berkeley Nuclear Laboratory Report* RD/B/N 4066
Speich G R and Schwoelbe A J 1975 *ASTM, STP*571 p40
Stephens R W B and Pollock A A 1971 *J. Acoust. Soc. Am.* **50** 904–10
Stone D E W and Dingwall P 1977 *Non-destr. Test. Int.* **10** 51–62
Tobias A 1976 *Non-destr. Test. Int.* **9** 9
Watkins P V C 1972 *Welding Institute Report No.* 3329/5/72
Wilson I and Fowler P 1973 *Proc. Symp. Nuclear Power Plant Control and Instrumentation, Prague* 1973 (Vienna: IAEA)
Wright C P and Washburn B 1976 *Proc. 22nd Int. Instrumentation Programme, San Diego* 1976 p235
Ying S P, Hamlin D R and Tenneberger D 1974 *J. Acoust Soc. Am.* **53**

3 Acoustic Emission Related to Metallurgical Effects

3.1 Acoustic emission and fracture mechanics

3.1.1 Experimental results

Owing to the importance of acoustic emission as a method of evaluating faults in pressurised structures, from the earliest days of the subject attempts have been made to correlate emission rates, total counts or emission energy with the severity of the faults causing the emission. Early work by Corle and Schliessmann (1972, 1973) shows how fracture mechanics may be used to normalise a number of acoustic emission readings derived from specimens containing flaws of differing sizes. Similar pioneering work was carried out by Dunegan and Harris. Figure 3.1 shows total counts against stress for three failures in a 4·5 mm thick, 51 mm wide, 4335B vacuum-melt carbon steel tensile specimen having different flaw sizes. It can be seen that emission counts increase rapidly above the flow stresses, indicated as s_1, s_2 and s_3, on the diagram, which vary with flaw size. However, if the counts are plotted against the stress intensity factor for the three specimens (as shown in figure 3.1) then a rapid increase in count occurs only in the region of the critical stress intensity factor which is independent of flaw size. While it is not the purpose of this book to review the science of fracture mechanics, and many standard texts have been written that provide a full account of this important subject, we will remind the reader that a crack of length a will grow according to the relation,

$$\sigma_c = K_{IC} \, (\pi \, a \, Q)^{-\frac{1}{2}} \qquad (3.1)$$

in which σ_c is the fracture stress. The crack will grow when the stress intensity K exceeds the critical stress intensity factor, K_{IC}. The geometrical factor Q has been tabulated for most geometries by Tada *et al* (1974). Classical fracture mechanics cannot be applied in its simplest

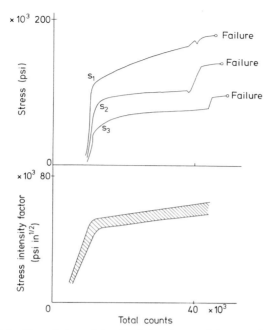

Figure 3.1 Use of stress intensity factor to correlate acoustic emission counts.

form to the design and study of pressure vessels as, strictly speaking, it is concerned with brittle materials whereas pressure vessels are thought to fail because of slow plastic growth. The concept of crack opening displacement (COD) was developed with this type of failure in mind. The K_{IC} and COD approaches are essentially complementary — as will be seen.

A paper by Elliott *et al* (1971) gives a good introduction to this concept. Figure 3.2 shows how COD relates to the 'classical' linear elastic fracture toughness (K_{IC}) approach. Where fracture occurs under essentially elastic conditions the plane strain fracture toughness, K_{IC}, is calculated from the force at the initial instability — point A in the figure. Above general yielding the linear elastic approach is evidently questionable and so COD comes into its own. When macroscopic ductile tearing becomes evident (past point B in the figure) the COD approach is less relevant. The COD values, defined as the plasticity controlled separation at the crack tip during loading, in a test of a particular steel, are obtained from displacement measurements made using a linear clip gauge — as shown in a paper by Fearnehough *et al* (1971). The value of COD taken to be indicative of the material's resistance to failure is that for the attainment of maximum force, i.e. point B in figure 3.2.

As pressure vessels usually fail in a slow ductile manner, Palmer, Holt and co-workers studied the laboratory behaviour of a typical pressure vessel steel under these conditions. Their classical results (Palmer 1973) are shown in figure 3.3. These important steels were also studied by

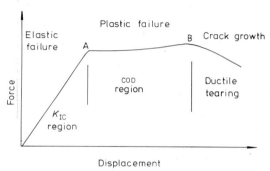

Figure 3.2 Appropriate regions for use of K_{IC} and COD criteria of failure.

Dunegan *et al* (1969) and Pickett *et al* (1971); Holt and Palmer have also studied the failure of a creep resistant steel and Palmer (1973) also lists the effects of annealing and corrosion on acoustic emission behaviour. In figure 3.3 the most important features to note are: (1) that the emissions occur up to just beyond general yield; and (2) that a considerable proportion of the ductile crack growth which occurs towards the end of the test is not accompanied by appreciable acoustic emission. Holt and his co-workers also point out that a pre-strain of 10% effectively removes acoustic activity, which is not recovered unless full metallurgical recovery by an appropriate heat treatment is obtained.

It is now generally held that the 'burst' type emissions detected are associated with the fracture of pearlite colonies. This is consistent with the correlation found by Palmer and Heald (1973) between the (calculated) plastic zone size and emission count rates. Supporting evidence

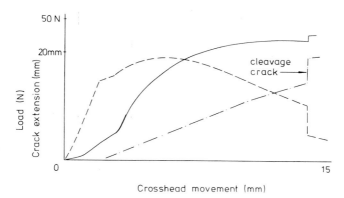

Figure 3.3 Total emission count (full curve), load (broken curve) and ductile crack extension (chain curve) for a C–Mn 'pressure vessel' steel.

is provided by the consistency of the acoustic emission amplitude spectrum during the course of the deformation, which suggests that the signals arise from a consistent range of fundamental processes rather than an effect which becomes severe as deformation proceeds. At elevated temperatures, in the region of 200 °C, emission associated with deformation of the matrix takes place. However, the following discussion will concentrate on room temperature tests carried out under slow ductile crack extension conditions. It should be noted though, that the work of Palmer and Heald is technologically significant as it indicates the strong possibility of detecting cracks associated with oxide formation at elevated temperatures, which includes reactor working temperatures.

An important study of the effect of steel strength on emissions has been carried out by MacIntyre and Green (private communication). The effects of proof stress, carbon content, sulphur content and cerium treatment on the acoustic emission associated with stable ductile cracking during fracture toughness testing have been evaluated in tests with experimental low alloy steels. During this work the increased level of mechanical background noise restricted attention to acoustic emissions of high amplitude. The greatest influence on emissions of this type is exerted by the proof stress of the steel. Over the range of 0·2% proof stress values (960–1530 N mm^{-2}) which was studied, a progressive increase in the extent of high-amplitude acoustic emission was found with increasing strength level. This is attributed to the effect of proof stress on the elastic strain and the degree of constraint within the plastic zone at the crack tip within which ductile cracking occurs.

A secondary influence of sulphur content on the amount of high-amplitude acoustic emission was observed for unmodified steels. In such materials the extent of acoustic emission was lower in steels with high sulphur contents. This effect is attributed to the reduction in local constraint associated with large elongated sulphide inclusions.

Masounave *et al* (1976) have also noted that: (a) the appearance of emissions, (b) the formation of a contraction zone at the crack tip, and (c) the deviation from linearity in the load extension curves occur at approximately the same time during a tensile test of 'fatigue'-produced cracks.

3.1.2 Models

It is instructive to consider a simple fundamental model of acoustic emission production during plastic yield in which a crack grows, pushing a plastically deformed zone immediately in front of the crack tip; this moves into the undeformed steel, producing microcracks across the colony as the region of general yielding proceeds. The observations of Palmer, Dunegan, Ying and their co-workers all indicate that the source of emissions is the immediate vicinity of the elastic–plastic interface. If we postulate that the total acoustic emission count, N, is directly proportional to the plastic zone *linear* dimension, s, then,

$$N = Ds \qquad (3.2)$$

where D is a constant of proportionality depending on strain rate, temperature, thickness and microstructure. Dugdale (1960) has shown that this linear dimension, s, for a wide centre-cracked plate pulled in uniaxial tension is

$$s = c \left[\sec(\sigma\pi/2\sigma_1) - 1\right] \qquad (3.3)$$

where c is the half crack length, σ is the applied stress and σ_1 is a stress characteristic of the strength of the material. Consequently,

$$N = Dc \left[\sec(\sigma\pi/2\sigma_1) - 1\right] \qquad (3.4)$$

If this equation is plotted on a logarithmic basis, it can be approximated by $N = \sigma^m$. Investigators have shown that m varies between 2 (in the case of results reported by Palmer and Heald 1973) up to 8. These results are entirely consistent with equation (3.4). Since, in linear fracture mechanics

$$\sigma_f = (2/\pi)\, \sigma_1 \sec^{-1}[\exp(\pi K_{IC}^2/8\,\sigma_1^2 c)], \tag{3.5}$$

it follows that

$$N_f = Dc\,[\exp(\pi K_{IC}^2/8\sigma_1^2 c) - 1] \tag{3.6}$$

(equation (9) of Palmer and Heald 1973). In the limit of small stresses both these equations reduce to

$$N_f = D\,\pi K_{IC}^2/8\sigma_1^2. \tag{3.7}$$

An earlier model, due to Dunegan *et al* (1969) is worthy of mention at this stage. In this the assumption is made that the total number of counts is proportional to the *volume* of the zone subjected to plastic deformation near the tip of the growing crack. The reader is referred to the original paper for a derivation of the relationship between counts and stress intensity factor, but Dunegan derived a relationship of the form $N = DK^4$, not the squared exponent of the model discussed above. It can be seen from the results shown in table 3.1 that in the present state of knowledge

Table 3.1 Values of n in $N = b\,\delta^n$

Source	Steel	Notch type	n
Arii *et al* (1975)	low C	machined	2·15
		sparked	0·97
	Q & T	machined	1·89
		sparked	3·48
	high C	machined	6·0
Palmer *et al* (1974)	C–Mn	pre-fatigued	1·0
Mirabile (1977)	C–Mn	pre-fatigued	4·0

it is impossible to reconcile these two models and the results obtained. It is consequently dangerous to read too much into the absolute determination of counts during pressure testing; rather one should rely on defect location coupled with conventional non-destructive examination of appropriate areas.

Owing to the difficulties and dangers of applying fracture mechanics to cases where general plastic yielding is taking place rather than brittle fracture, the measurement of crack opening displacement (described above and developed by Walker and others) is normally used to characterise the 'plastic' metal failure usually met in pressure vessel and pipeline steels. Much work has been carried out to correlate crack opening displacement measurements with acoustic emission measurements in order (it is hoped) to enable emissions to characterise the severity of cracks in structures. Work has been described by Arii *et al* (1975), by Mirabile (1977) and by Palmer *et al* (1974). We follow the treatment of Palmer *et al* to provide a useful framework for subsequent discussions.

From the Bilby model (Bilby *et al* 1963) of plastic relaxation around the crack, the crack tip opening displacement (see Elliot *et al* 1971) is related to the crack length, a, and a plastic zone size, s, by

$$\delta = [4(1 - \nu)\sigma_1 a/\pi\mu] \ln [1 + (s/a)] \tag{3.8}$$

where μ is the shear modulus and ν is Poisson's ratio. If we assume that the emission rate is proportional to plastic volume, then $N = Ds$, and so

$$N = Da \{\exp[\pi\mu\delta/4(1 - \nu)\sigma_1 a] - 1\}, \tag{3.9}$$

from which, for small values of the exponents, we obtain

$$N = D\pi\mu\delta/4(1 - \nu)\sigma_1. \tag{3.10}$$

A plot of emission count against COD quoted by Palmer *et al* (1974) is shown in figure 3.4. These results are similar to those obtained by Bentley (1974) but differ from other workers. Table 3.1 shows the exponents of the relationship between total counts and COD as measured by different authors. The difference between the results of Mirabile (also shown in figure 3.5) and those of Palmer are particularly significant as the materials tested are similar. At the time of writing no explanation can be found for this large variation in experimental results. It is interesting to note however that the comprehensive paper by Mirabile also confirms the hypothesis that emissions in this mode of fracture (slow ductile crack growth) occur when cracks are formed in the pearlite colony in the elastic–plastic interface. The work of Masounave *et al* (1976) referred to above shows that the simple models of Dunegan *et*

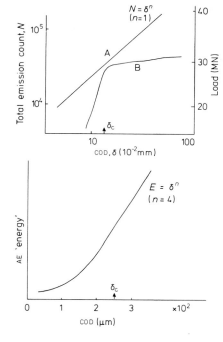

Figure 3.4 Plots of count (A) against COD, and load (B) against COD, for a C–Mn 'pressure vessel' steel (Palmer *et al* 1974).

Figure 3.5 Plot of acoustic emission 'energy' against COD for a C–Mn steel (Mirabile 1977).

al (1969) do not provide a realistic basis for correlating acoustic emission with stress conditions.

3.1.3 Codes

The American Society for Testing and Materials and the American Society of Mechanical Engineers have both produced draft codes of practice for carrying out acoustic emission testing on pressurised components. The ASME code is of widespread interest because, in the United States, these codes can become mandatory for insurance and certification purposes. This code is straightforward and simple and concentrates on the location of defects rather than the evaluation of defect severity, and calls for the use of an impulse tester to calibrate location systems.

In contrast the ASTM code attempts to give some credence to defect evaluation by categorising the variation of total counts with applied pressure on load into three distinct categories, A, B and C, depending on the shape of the curve. In my opinion it is not meaningful to categorise the results of acoustic emission tests in this way at this stage of our knowledge of the behaviour of materials under loading conditions.

3.2   Acoustic emission under fatigue conditions

It has been pointed out that for the in-service testing of many structures, including pressurised components and offshore structures, a knowledge of acoustic emission of materials under fatigue conditions is essential. This has been emphasised (Smith and Warwick 1974) in connection with pressurised plant, but the need to monitor structural integrity, especially of aircraft and of offshore platforms, makes acoustic emission monitoring of structures under fatigue-producing conditions doubly important. While some of the earliest promising acoustic emission investigations were carried out on the fatigue behaviour of metals, it is still true that

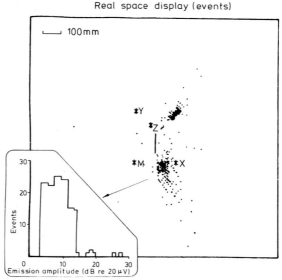

Figure 3.6   Emission from a partial penetration crack during fatigue cycling.

the growth of fatigue cracks can be reliably monitored but that it is not yet possible to estimate life before fatigue failure. Early work concentrated on correlation between the counts received for each fatigue cycle and the length of fatigue crack growth; later results show that it is important to understand how emission intensity varies during each cycle. Whilst the earliest simple models aimed to explain emission rates during fatigue by the proposition that emission rate varies in proportion to the energy of new metal surfaces formed in fatigue crack growth, this can no longer be regarded as valid, as current work on mechanisms leading to emission during fatigue crack growth is rapidly gaining insight into the relationships between growth rates and emission parameters.

Examples of work carried out in this field are given in the papers by Dunegan *et al* (1970), Harris (1974), Harris and Dunegan (1974), Hartbower *et al* (1973), Lindley *et al* (1976), Morton *et al* (1973), Nakasa (1973) and Sinclair *et al* (1976, 1977). Sinclair (1977) has also written a useful brief summary of the current state of acoustic emission for in-service non-destructive examination of structures and components under fatigue conditions. He shows how emissions are detectable in operating plant subjected to cyclic loading using the ACEMAN system (see chapter 2). Emissions were generated from fatigue cycling of a 6 mm wall mild steel pipe containing an axial partial thickness slit. The display, shown in figure 3.6 covers the 1400 load cycles commencing 4500 cycles before final failure.

The simple relationship

$$dN/dn = C_1 \Delta K^m, \qquad (3.11)$$

where N is the total number of counts, n is the number of cycles, ΔK is the range of K, and C_1 and m are constants, has been shown to apply during tests in which the stress intensity factor amplitude has been varied. The rate of crack length growth per cycle follows the same equation with, in some cases, the same exponent value, i.e.

$$da/dn = C_2 \Delta K^{m_1} \qquad (3.12)$$

($m = m_1$ in many cases).

Webborn (private communication) and other workers have studied the way in which the counts per cycle vary with the energy released per cycle. Figure 3.7 shows a compilation by Webborn of this parameter for various materials. The energy released per cycle is given by equation

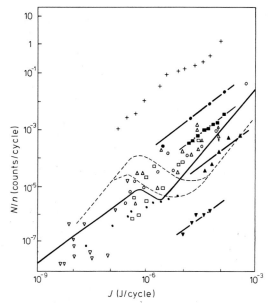

Figure 3.7 Counts per cycle, N/n, as a function of energy released per cycle, J, for various materials: ——, 7075 T6 Al; ---, 4140 steel; $\triangle$, 2024 T851 Al; $\bigcirc$, 2024 T851 Al; $+$, EZ33A T5 Mg; $\square$, Ti; $\triangledown$, IN100 Ni; (all data from Harris, 1974): $\bullet$, 1% C steel; $\blacksquare$, DUCOL W30B; $\blacktriangle$, EN 30B; $\blacktriangledown$, C–Mn steel; (all data from C J H Webborn, personal communication).

(3.13) below and is calculated on the assumption that energy released is derived from expansion in the volume of the ductile region at the head of the crack:

$$J = (B/E)[\Delta K/(1 - R)]^2 (da/dn) \qquad (3.13)$$

$R = K_{min}/K_{max}$.

The logarithmic relationship between counts per cycle and stress intensity range has been observed by many authors, Sinclair, Webborn, Harris and Dunegan included. Figure 3.8 is from Sinclair's work. The values of the exponent and of the constant of proportionality are a function of the material and of the conditions under which the material is tested. Sinclair *et al* (1977) point out that the implication of figure 3.8 is that the total number of events during cyclic loading in which a

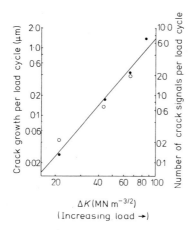

Figure 3.8 Crack growth and emission count rate during fatigue cycling of A533B steel: ●, crack advance per cycle, da/dn; ○, events per cycle, dN/dn; full curve plots ΔK^m for $m = 2.5$.

fatigue crack area A is created, is directly proportional to that area, i.e.

$$N = \gamma A. \qquad (3.14)$$

The constant in the above expression is the specific emission activity for fatigue crack growth in the material under test. If it has a high value then the material will allow ready detection of fatigue cracks. It is not possible to give a universal figure for the given material as the value also depends upon the atmosphere in which the crack is formed. Over the range of crack propagation rates for which acoustic emission was measured in an A533B steel by Sinclair and his co-workers the activity parameter was 44 events per square millimetre of crack area created. Bartle uses a similar specific activity figure but it is not possible to compare Bartle's figures with those of Sinclair and others because the equipment used for counting was of a different nature.

Sinclair also carried out tests also using low-carbon pipe steel (Sinclair *et al* 1977). In this case the number of events per cycle has a relationship of the form of equation (3.11) to the amplitude of the stress intensity factor but the exponent was 4·0 rather than 2·5 for the A533 carbon manganese steel (the low-carbon pipe steel had a carbon value less than 0·25%).

It is possible at this stage in our knowledge of acoustic emission from fatigue crack growth to make some deductions concerning the mechanisms of emission. These have been treated systematically by Sinclair *et al* (1977) and by Bartle (1979) and Lottermoser (1979). The three mechanisms possible are: (1) new yielding at the edge of the plastic zone (known to produce emissions during the first loading); (2) microfracture processes in the region of intense plastic strain close to or at the crack tip (this mechanism has also been postulated for single loading by Brindley and Harrison (1972) and also by Masounave *et al* (1976)); (3) purely mechanical processes near the tip of the fatigue crack, referred to by Sinclair and others as 'unsticking'. It should be possible in principle to determine the type of mechanism concerned using location equipment but in practice the areas concerned are so close to each other that this is not possible.

If we consider the first mechanism we would imagine that the activity would be related to the volume of the plastic zone. It has been shown by Sinclair and many other workers that, for this mechanism, the number of emissions is directly proportional to the volume of the plastic zone, i.e. $N = BV_p$. Now the size of the plastic zone per head of the defect is given by

$$r_p = (1/3\pi)(K/\sigma)^2 \tag{3.15}$$

with σ being yield stress. This has been deduced by Tada *et al* (1974). As we know that, during cycling at a constant load, the plastic zone both moves with the crack and grows in size, we can readily show that

$$dn_p/da = (1/3\pi a)(K/\sigma)^2 \tag{3.16}$$

(ignoring the effect of change in specimen geometry with increasing crack length). Consequently if the plastic zone is a cylinder with diameter r_p and length B (equal to specimen thickness) we obtain

$$dV_p/dn = V_p B(da/dn). \tag{3.17}$$

Consequently we find that

$$dN/dn = (C_2 B\beta/3\pi\sigma^2)[\Delta K^{4.5}/(1 - R)^2] \tag{3.18}$$

($R = K_{min}/K_{max}$). Since r_p is approximately constant the rate of emissions per cycle is an exponential function of the change in stress intensity factor to the power of 4·5. It can be seen that this is contrary to experience except in the case of low carbon steel.

The relationship deduced in the classical work by Dunegan *et al* (1970) is similar to that given above. The basic assumptions are similar, with the additional assumption that crack growth occurs at the maximum load. It can be seen that the relationships, deduced from the assumption that emissions occur from the growth of the plastic zone and are produced by new yielding at the edge of the plastic zone, are very unlikely to be realistic for all materials.

In addition the emission activity predicted by equation (3.18) (taking a reasonable assumption for the activity parameter B) is much less than observed (this equation has been the basis of several gloomy prognostications concerning the ability of acoustic emission to detect the growth of cracks under service conditions and has led to much confusion in the recent past). Sinclair and others take β as 2×10^{-3} emissions/mm^3, so if ΔK is 47·3 N m$^{-3/2}$ we obtain the ratio of observed to predicted emissions as high as 10 000. Even if we assume different β values we are unlikely to match real with theoretically predicted emissions using equation (3.18). Furthermore Sinclair and his co-workers have noted that emission activity continues following a reduction in peak stress intensity factor. In the case of fatigue cycling at K_{max} equal to 35·5 MN m$^{-3/2}$ the plastic zone calculated from equation (3.18) did not pass outside the zone previously established in the prior cycling at 89·4 N m$^{-3/2}$.

It is more difficult to differentiate between the mechanisms (2) and (3); current thinking is that both occur. One way of examining this question is to note when emissions occur during a cycle. Mechanism (3) should occur early in the cycle whereas mechanism (2) should give more numerous emissions near to the maximum load. Some evidence concerning this has been produced by Bartle (1979), and the work of Lottermoser (1979) at Saarbrücken has also thrown light on this factor.

The third mechanism described above should be dependent on the atmosphere in which the fatigue test is carried out. Early work by Hartbower *et al* (1973) illustrates the dramatic effect that the media normally associated with stress fatigue conditions can have on the number of emissions. Figure 3.9(*a*) shows results from a specimen of titanium tested in air and tested in air and water. It can be seen how the number of counts changes at the same K_{max} values. Figure 3.9(*b*) also shows one experiment where water was added during the test; the number of

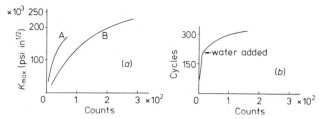

Figure 3.9   Effect of atmosphere/environment of test on acoustic emission count. (*a*) Test in air, A, and water, B. (*b*) K ≃ 820 MPa (119 × 10³ psi).

emissions increases dramatically. This is of course associated with the increased crack growth rate.

The recent work by Lottermoser and by Bartle has shown why, under some circumstances, the emissions counted per cycle from a fatiguing crack are more numerous than expected from the simple ductile crack growth theory. The way in which emissions occur during each loading and unloading cycle must be examined. Bartle's (1979) work has shown how high-amplitude emissions occur near the minimum of the loading curve in tests on structural steel and similar behaviour has been observed by Lottermoser. Figure 3.10 shows an interesting curve of emission energy for different parts of the loading and unloading cycle as a function of the number of cycles during the test run. It is seen from this latter

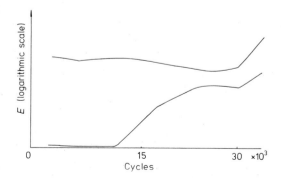

Figure 3.10   Acoustic emission energy per load cycle, *E*, as a function of number of load cycles; upper curve at maximum of load cycle, lower curve at minimum.

figure that the lower loading curve is indicative of the region where the crack is growing under fatigue loading conditions. It is now thought that the high-amplitude emissions during the low loading part of the cycle are due to rubbing of crack faces, i.e. mechanism (3) above. This type of emission is very valuable for the detection of growing cracks in fatiguing structures and it is thought that some of the results where high-amplitude emissions have been observed in offshore structures reported in later chapters can be explained in this way. My own view is that the incidence of high-amplitude, frequent counts at high loads in the later part of the life of a fatigue test piece can explain the curious 'S curve' behaviour found by Harris and Dunegan. It is postulated that the lower parts of the curve (characteristic of low growth rates) are due to plastic deformation, a comparatively quiet process. But when the crack growth rate is higher and/or the amplitudes of stress intensity variations are higher, then the rubbing factor takes over, producing the second and more noisy part of the curve.

3.3 Summary

The results of the major investigations of acoustic emission during 'ductile' crack growth and during fatigue (both in metals) have shown that:

(1) Acoustic emission occurs during 'ductile' crack growth in pressure vessel manganese steels. In room-temperature laboratory tests most of the emissions occur up to yield.

(2) High-temperature slow ductile crack growth tests in similar steels show that acoustic emission could be used for 'in-service' structural integrity monitoring of boiler plant and in nuclear reactors.

(3) The rate of emission in steels, under similar environmental conditions, varies with yield of the material, increasing as the yield point increases.

(4) Fatigue testing of steel is accompanied by acoustic emission — thought to be due to mechanical as well as metallurgical effects. Acoustic emission can be used to detect the presence and position of growing fatigue and stress – corrosion fatigue cracks; our understanding of the processes involved do not as yet allow reliable estimates to be made of the remaining life of a structure.

(5) The environment (e.g. salt water or hydrogen-rich gas) in which
     acoustic emission tests are carried out is a vital factor in determining
     the rate of emissions from a steady crack growth or a fatigue crack
     growth situation.

Great care must be taken in simulation studies of acoustic emission to
note the effects of environment. It is easy to be excessively optimistic
or pessimistic when estimating the usefulness of acoustic emission in
monitoring a structure from laboratory tests.

References

Arii M, Kashiwaya H and Yanuki T 1975 *Engng Fract. Mech.* **7** 551
Bartle P 1979 *Final Report ECSC Project* 7210 GA8803
Bentley P 1974 *Acoustic Emission and Pressure Vessel Failure, Inst.
    Mech. Engineers (London) Conf. Publicn 8*
Bilby B A, Cottrell A H and Swinden K H 1963 *Proc. R. Soc.* A **272**
    304
Brindley B J and Harrison R P 1972 *CEGB Report No.* RD/L/N137/2
Connors D C 1973 *CEGB Report No.* RD/B/N2626
Corle R R and Schliessman J A 1972 *Fall Conf. American Non-Destruc-
    tive Testing Soc., Cleveland, Ohio, 1972; Mater. Eval.* **30** 30A
——1973 *Mater. Eval.* **31** 115–20
Dugdale J 1960 *J. Mech. Phys. Solids* **8** 100
Dunegan H L, Harris D O and Tatro C A 1969 *Engng Fract. Mech.* **1**
    105
Dunegan H L, Harris D O and Tetelman A S 1970 *Mater. Eval.* **28**
    221–7
Elliott D, Walber E F and May M J 1971 *Inst. Mech. Engineers (London)
    Paper* C77/71
Fearnehough G D, Lees G M, Lowes J M and Weimer R T 1971 *Inst.
    Mech. Engineers (London) Paper* C33/71
Harris D O 1974 *Dunegan Endevco Report* DE-74-4
Harris D O and Dunegan H L 1974 *Expl Mech.* **14** 71
Hartbower C E, Marais C F, Reuter W E and Crimmins P O 1973
    Engng Fract. Mech. **5** 765
Lindley T C, Palmer I G and Richards C E 1976 *CEGB Report No.*
    RD/L/N134/76
Lottermoser 1979 *Final Report ECSC Project* 7210 GA1112

Masounave J, Lanteigne J, Baffin M and Hay D 1976 *Engng Fract. Mech.* **8** 701–9

Mirabile M 1977 *Proc. Conf. Fracture Mechanics, Hong Kong 1977*

Morton T M, Harrington R M and Bjeletich J G 1973 *Engng Fract. Mech.* **5** 691

Nakasa H 1973 *Proc. Symp. Nuclear Power Plant Control and Instrumentation, Prague 1973* (Vienna: IAEA)

Palmer I G 1973 *Mater. Sci. Engng* **1** 227–36

Palmer I G, Brindley B J and Harrison R P 1974 *Mater. Sci. Engng* **14** 3–6

Palmer I G and Heald P T 1973 *Mater. Sci. Engng* **11** 227–36

Pickett A G, Reinhart E R and Ying S P 1971 *Proc. 8th Symp. Nondestructive Testing, San Antonio, Texas 1971*

Sinclair A C E 1977 *CEGB Report No.* RD/B/N4066

Sinclair A C E, Formby C L and Connors D C 1976 *Acoustic Emission* ed R W Nichols (Barking, Essex: Applied Science) pp 51–63

—— 1977 *Mater. Sci. Engng* **28** 263–73

Smith T A and Warwick R G 1974 *Int. J. Pressure Vessels and Piping* **2** 283

Tada J, Payis P C and Irwin G R 1974 *The Stress Analysis of Cracks Handbook*

4 Application of Acoustic Emission to Pressurised Components

4.1 Introduction

The testing of pressurised components and plant was the earliest industrial application of acoustic emission, and it remains the most advanced and important area of the technique. Proof testing and in-service monitoring of the integrity of nuclear generating plant and chemical manufacturing pressure vessels provided early impetus to its development.

It is of interest to note that many of the workers involved in the initial development of acoustic emission were concerned with engineering aspects of nuclear power generation. As a consequence, this chapter describes those aspects of acoustic emission which have received the most intensive study over the past ten years and which, arguably, have the most thorough basis in theory and understanding of the results which can be obtained using the technique.

This chapter will describe the application of acoustic emission to the hydrostatic testing of pressure vessels before service, the in-service monitoring of vessel integrity, the detection of stress–corrosion induced faults in nuclear power plant, and the testing of pressurised gas transmission lines. It will also discuss what is probably the most successful new development of the technique, the detection of leaks in cooling systems. The state of progress made in these areas may be summarised as follows.

Hydrostatic tests on pressure vessels. Acoustic emission is becoming accepted as a viable method for testing these critical components before service and may become obligatory in the United States and other countries having advanced nuclear power generation programmes. It has in one case detected faults not disclosed by conventional NDT methods. It cannot however be used as the sole method of testing.

52

In-service testing of pressure vessels. Because of the difficulty of obtaining access to pressure vessels whilst in normal service, acoustic emission is used to test the integrity of vessels of nuclear reactors whilst in service. The tests which will be described in this chapter have shown that it can detect in-service faults with sufficient certainty to justify an expensive plant shutdown and full non-destructive examination of the pressure vessel. Acoustic emission in-service testing is now regarded by some authorities (in the USA, West Germany, Sweden and Finland, but not as yet in the United Kingdom) as a fully proven technique to check the integrity of pressure vessels, and is proving to be very valuable in other areas, notably in offshore structures where access for 'follow-up' inspection does not require shutdown of the plant.

Surveillance of components subjected to stress–corrosion. In this special aspect of 'in-service' testing, where stress–corrosion is a major factor in component failure, acoustic emission has been shown to be a valuable tool, and work in the United States will be described below which demonstrates this.

High-pressure gas pipelines. Opinions vary about the usefulness of acoustic emission for the detection of faults in high-pressure pipelines. There is no doubt that if faults occur near welds, emissions can be detected and the faults located. However, in some cases failure or defects occur due to mishandling (during construction or subsequent digging operations) in parts of the line between the welded areas. In these cases, it has now been demonstrated that faults can be detected with some degree of certainty by acoustic emission techniques.

Leak detection. This exciting new application of acoustic emission is proving to be most valuable in leak detection in nuclear reactors, chemical plant, iron and steelmaking plant and in pipelines.

These various applications will now be discussed in detail.

4.2 Hydrostatic test monitoring of pressure vessels

Acoustic emission tests carried out on pressure vessels up to the time of writing are summarised in table 4.1. The table includes tests carried out by Parry on behalf of Exxon Research and Engineering Corporation

Table 4.1 List of pressure vessel proof test applications

Item	Where and/or why test was performed	Material	Wall thickness (cm)	Diameter (m)	Length (m)	Number of vessels tested	Identity of emission sources	Tested by	Reference
German nuclear plant pressure vessel and hydraulic components	Requalification hydro-tests			1·6		1	50 mm long defect in plate weld defect	(Exxon) (Parry)	Parry (1975)
Experimental vessel	Culcheth	BS1501-224-32A (normalised)	25	1·6	4·3	1	Axial path through artificial flaws	Birchon Parry	
Experimental vessels		Low alloy steel	7·6	1·5	5·5	2	2 cm deep cut	Bentley	
HSST (Heavy Section Intermediate Steel Test) vessel	Nuclear reactor vessels, (Oak Ridge National Laboratory)	A-508	15·2	1		6	No significant emitters — confined by conventional NDT	Kelley Parry	
Redundant pressure vessel (EBOR)	Experimental programme lasting 1 yr — including in-service tests (Edison Electric/TVA)					1	Artificial defects and welds	Exxon South West Research Institute Bell	
Reactor pressure vessel (LaSalle II) (BWR)	First hydro-test (EPRI programme)					1	41 AE sites — well correlated with ultrasonically detected defects.	EPRI (Stahlkopf)	Stahlkopf and Green (1976)

Component	Application	Material				No.	Comments	Reference	
Pressure vessel (BWR) (Also nozzles and pipes)	Hydro-tests at CAORSO (Italy)	ASME SA533	136	5·87		1	One main event on a pipe—confirmed ultrasonically	ENEE (Rome)	Watanabe et al (1976)
Pressure vessel	Experimental hydro-test	SB 42 and SB 49·M	22	1·5		1	8 cracks, 2 slag pockets, 2 blow holes, 2 poor welds	Watanabe et al	
Header	Test to destruction under pressure	Mild steel					Much activity prior to failure	Palmer	
Petrochemical reactor	Concurrent with acceptance hydro-test of new vessel	A-204 Gr. C	5·08	Spherical	7·23	2	Weld porosity and slag inclusions undercutting in welds attaching bracket; no repairs required	Exxon	
Product storage sphere	Concurrent with acceptance hydro-test of new vessel	A-516 Gr. 70	6·0	Spherical	13·72	2	Small cracks in welds attaching clips and brackets; weld porosity; no repairs required	Exxon	
Petrochemical reactors	Concurrent with acceptance hydro-test of new vessel	A-302 Gr. A	10·16	4·88	15·84	1	Weld porosity and slag inclusions not requiring repair	Exxon	
				4·88	32·00	1	Cracks in weld-overlay cladding lining which were repaired		

Acoustic Emission

Item	Where and/or why test was performed	Material	Wall thickness (cm)	Diameter (m)	Length (m)	Number of vessels tested	Identity of emission sources	Tested by	Reference
Petrochemical reactor (Culcheth)	To requalify for continued service Concurrent with hydro-test	A-201 Gr. B	6·67	3·66	2·74	1	Two weld undercuts in lug attachment welds ground out with surrounding plate; crack in bottom of gouged area on shell ID ground out and blended in with adjacent surface 1 in diameter lamination in plate; no repair	Exxon	
Petrochemical reactor	To requalify for continued service Concurrent with hydro-test	A-212	4·76	3·58	2·66	1	Plate lamination 1 in. diameter; no repairs required	Exxon	
Gate valves	To requalify for continued service Concurrent with hydro-test	Cast 1¼ Cr-½ Mo	5·8-7·6	16 in ring joint, 600 psi rating		1	Large surface crack 3 in long by ¼ in wide on ID by visual inspection; also internal shrinkage; cavity found by radiography	Exxon	

Pipeline	R & D concurrent with hydro-test	Seamless carbon steel	Schedule 40·55	7·6–10·16	14 in ring joint, 600 psi rating 3 in NPS	One section 21,400 mm long	Internal shrinkage cavities found by radiography Many surface cracks (up to 1 in long) on ID	This section of pipeline contained an artificial flaw in which failure occurred; this and other sites identified during acoustic emission; minor emitters were from random inclusion in pipe wall	Exxon
KEMA vessel	Experimental tests	Low alloy steel ($K_2 = 184 \times 10^3$ psi in)				1	Artificial defects at welds	Kelley
HSST vessel	Oak Ridge National Laboratory			1·5		2	Artificial defects and welds	Gopal *et al* Gopal *et al* (1976)
Experimental tests	CSM Rome	C–Mn	~0·5	1·5	1·0	12		Mirabile

and those tested by Stahlkopf and Green (1976). The results shown in the table indicate that acoustic emission is a most useful technique for monitoring hydro-tests especially when a location system is also employed in the test. Hydrostatic testing is normally carried out under circumstances where the availability of normal and conventional methods such as radiation, ultrasonic and other techniques is assured. Consequently the purpose of the acoustic emission tests is to complement conventional NDT methods. It is unlikely that acoustic emission would ever replace radiography or ultrasonics, but it does provide a very useful additional tool for examining large structures. Transducers do not have to be nearer than a few metres to a fault area, so full coverage of a large vessel is easy; continuous monitoring is possible without the need for supervision of equipment, and the method lends itself to modern electronic techniques to aid the interpretation of the output signals.

Probably the most comprehensive and objective description of pressure vessel testing using acoustic emission is that of Birchon *et al* (1974), which will be taken as a model for the numerous other descriptions published. These workers used the particularly well designed low-noise (AML) system described in chapter 2. The vessel itself was 1·6 mm in diameter, 4·3 m long and made from 25 cm thick plate of BS 1501-224-32A normalised steel. The vessel had been specially prepared by welding in an insert panel having four axial partial thickness defects, each 200 cm long, cut in the vessel. The defects were spark eroded to 20 mm and 17·5 mm depth and the vessel cycled from 0 to 3·62 MN m^{-2} (0–525 psi) to develop fatigue cracks at the base of each defect until the depth had been increased to about 20 mm.

In addition to the deliberate defects four mild steel EN 4A bars, each 20 mm square by 900 mm long, were welded tangentially to the vessel near one end and notched close to the vessel surface so that, as the unsupported end of the bar was oscillated, a crack would be initiated and propagated through the weld metal and heat-affected zone parallel and very close to the surface of the vessel. In this test, developed by the South West Research Institute (USA), the sides of the bars were polished to permit the rate of crack growth to be monitored. An interesting extra item was the fixing of an 'AML emitter' (see chapter 2) to the vessel as a controllable source of stress wave emission, comprising a double cantilever beam specimen of a high-strength aluminium alloy attached to the structure using an adhesive. One end of the aluminium bar is notched and the notch can be forced apart by screws; thus tightening the screws propagates a crack from the root of the notch. Stress–corrosion cracking, initiated by a drop of 3% salt solution added to the notch, produces copious emissions, random in timing and energy level.

The vessel, which was maintained at 45 °C throughout the test, was monitored with a total of 16 PZ transducers having a resonance frequency (450 kHz) chosen to avoid interference from hydraulic noise. The transducers were placed irregularly on the vessel with one transducer near to each of the artificial defects and others about 300 to 600 mm away. The test vessels were pressurised a total of 13 times, to a maximum pressure on the monitoring cycle of 6·55 MN m^{-2}. Failure occurred on the fourth high-pressure cycle.

Birchon concluded from these tests that a 94% probability exists that a location plotted by an acoustic emission defect system would be confirmed to contain a defect by subsequent ultrasonic inspection. Strictly speaking, this applies only to the AML defect location system but it is unlikely that other systems would produce greatly dissimilar results. Variation of the system threshold level gives a useful sensitivity control and reduces the number of 'rogue' readings. The difficulty, found by Birchon, of detecting slow crack growth from artificially induced defects has also been found by many other workers.

Acoustic emission has the great advantage that activity is related to the actual dangers involved; this is in contrast to other forms of non-destructive testing which can detect the presence of defects with facility almost inversely related to the dangers which such defects may pose to the structure. An important set of results were also obtained from the requalification hydrostatic tests on the KRB nuclear power plant in West Germany (Parry 1975). Twelve emission sources were located during the test, nine of which were judged to be minor. The remaining three sources were inspected by ultrasonic techniques under the direction of the German Safety Inspection Authority, indicating that one source was a 50 mm long defect located 60 mm below the outside diameter surface. A second source, found in the vessel circumferential valve, showed under ultrasonic examination two closely spaced defects, one 20 mm and the other 10 mm long. The location accuracy varied from ±20 cm on piping to ±3 cm on pumps, with location accuracy of ±7 cm on the vessel itself. Similar results were also found by Gopal *et al* (1976) for a Westinghouse light water reactor system (nuclear power generating equipment installed on a world-wide basis). Special tests were carried out for the Oak Ridge National Laboratory on two 1 m (39 inch) outside diameter steel pressure vessels using 15 cm (6 inch) thick plate, containing prepared and sharpened surface cracks. Figure 4.1 shows the cumulative counts of one test compared with the circumferential strain in line with the emitting flaw. This plot illustrates a good correlation of the acoustic counts with the strain data. The increased count level corresponds to the transition of strain level from the elastic to plastic range and shows how the counts rate increases near to the failure.

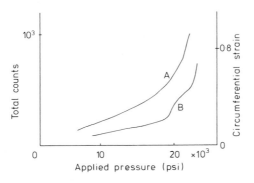

Figure 4.1 Total acoustic emission counts, A, plotted with strain measurements, B, for a pressure vessel test.

Of the many tests given in table 4.1 the Japanese experiment, reviewed by Watanabe (1976) is particularly well reported. Good correlation was found between emission and the behaviour of flaws in the vessel during a hydro-pressure test, including a good correlation between crack length and source activity for a number of well located emission events. Although the fracture tests were limited in number the Japanese results showed that dangerous defects which grow and expand in pressure vessels under normal testing can be clearly detected by acoustic emission. Stable defects did not emit signals regardless of their size.

De Michelis and Mirabile (1978) have carried out recent and thorough tests in Italy, on a number of small special vessels in C–Mn steels under the auspices of the European Coal and Steel Community. In this work a total of twelve, small pressure vessels, especially made for the series of experiments, were prepared with artificial flaws, in general within the parent metal and not in welds. The pressure vessels were of two grades of steel (0·4% C, 0·75% Mn and 0·3% C, 1·18% Mn). The vessels themselves were approximately 1 m in length and 1½ m in diameter. One grade was water quenched from 850 °C and tempered at

550 °C. Up to three transducers were placed on the vessels and coupled to Fridet acoustic emission electronics. The transducers were Dunegan Endevco S140B resonant at 140 kHz and were all of comparable sensitivity. The experiments consisted in increasing pressure until leaks occurred. All vessels were equipped with a clip gauge for measuring COD as well as the acoustic emission parameters. The acoustic emission parameter of particular interest was the sum of the amplitudes of the impulses squared.

The final conclusions drawn by de Michelis and Mirabile were as follows:

(1) Experimental failure stresses agree satisfactorily with the values predicted by fracture mechanics for fractures induced by surface flaws, provided that the flow stress failure criterion is modified in such a way as to take into account the crack depth as the relevant dimension of the flaw rather than crack length. Conclusions are drawn concerning the question whether fracture occurs with leaks or whether rupture occurs before leak.

(2) Acoustic emission activity simultaneously recorded by three transducers located around the surface notch has shown essentially two patterns, as mentioned above; the first characterised by a small number of high-energy counts and the other by a larger number of low-energy counts. Electron microscopy has confirmed that the first is associated with cleavage cracks, probably in the pearlite colony, and the second in terms of simple ductile fracture mechanisms. No systematic correlation has been found between acoustic emission parameters and the mode of failure.

(3) Large differences have been observed in some cases in the absolute values of the event counts between transducers of the same triplet for the same test. This is not yet understood.

4.2.1 General conclusions concerning pressure vessel testing

The list of acoustic emission monitored hydrostatic tests on pressure vessels in table 4.1 must be regarded as impressive evidence of the value of this method for checking the safety of these important structures, where the consequences of failure are severe. As a result, in many countries acoustic emission monitoring is regarded as an almost essential adjunct to the use of ultrasonic and radiographic testing. In West Germany one case was reported where the conventional methods missed a serious case of lamellar tearing; acoustic emission alone detected the fault. However, other workers, especially in the United Kingdom believe that as a result of the nature of the emission–load characteristics of the low-alloy steels used (see figure 3.3 where most emissions occur well before the region of extensive crack growth), acoustic emission is of limited value in detecting faults in these steels used in the 'clean' conditions of pressure vessels. This is in contrast, it must be noted, to when such steels are used under conditions where stress–corrosion can occur.

I feel that acoustic emission should be regarded as a complementary tool — possibly an essential adjunct — to conventional methods for hydrostatic pressure vessel testing. This general conclusion does not extend to the inspection of autoclaves, where I have received reports of successful and accident-preventive applications of acoustic methods.

4.3   In-service surveillance

The use of acoustic emission for in-service surveillance is encouraging in some respects and discouraging in others. Holt and Palmer (1974a, b) in an early study of application of acoustic emission pressure vessel testing have shown that at service temperatures (100–200 °C) the microstructure of pressure vessel steel is such that emissions should regularly occur in fault areas. The problem of 'quiet steels' was not encountered by Holt and Palmer because of the occurrence of emission sources associated with deformation of the matrix of the steels used. Oxides also provide a source at working temperatures in many instances — see the section later on 'stress–corrosion'. Care has always to be taken with background noise in these applications. Recovery of acoustic emission does not occur unless metallurgical recovery or strain takes place and so 5–10% overpressurisation might be required to detect crack growth.

The usefulness of acoustic emission for the detection of leaks in power plants has now been well estalished. Gopal *et al*, as early as 1976, had documented this view, with the reservation that the detection of acoustic emission from certain inaccessible regions of the pressure boundary could not always be guaranteed at that time. Since then the developments of acoustic waveguides to reach inaccessible regions, coupled with the new signal validation techniques described in chapter 2, have eliminated these earlier difficulties. Originally the range of the system was limited

Table 4.2 Permanent installations.

Type of installation	Location of installation	Equipment maker
Wind tunnel pressure bottles	NASA, Langley	Dunegan Endevco
Chemical plant leaks	Dow Chemical, Plaquemine	Dunegan Endevco
KC135 tanker airplane wing box	Tinker Airforce Base, USAF	Dunegan Endevco
Reactor pressure vessel and by-pass valves	Chicago (Commonwealth Edison)	Dunegan Endevco
Input nozzle on nuclear reactor vessel	Philadelphia Electric	Trodyne
Autoclave	Lawrence Livermore Laboratories	Dunegan Endevco
Stainless steel piping	General Electric, San Jose	AET
Nuclear reactor	Japan	South West Research
Blast furnaces	Redcar, UK	Unit Inspection

to the upper vessel but now monitoring installations can and do completely cover the whole plant. A list of permanent, acoustic emission monitoring installations, including nuclear plant, chemical works and aircraft, is given in table 4.2.

4.4 Stress–corrosion cracks

Stress–corrosion cracking is responsible for many failures which occur in aircraft, chemical plants, power generation equipment, the oil industry and applications where steel and other metals come into contact with aggressive environments. Acoustic emission techniques offer the possibility of locating and detecting the growth of most cracks produced under these conditions. McIntyre and Green (private communication) in the United Kingdom have done valuable background work in this area.

One of the most promising applications of acoustic emission is the detection of stress–corrosion in power plant piping. This has been demonstrated by Stahlkopf *et al* (1976) in work carried out for the Electrical Research Institute in the United States. In this test facility work, a 10·16 cm (4 in) diameter light water reactor pipe of type 304 stainless steel was subjected to four-point loading in a loop of high-purity water maintained at a temperature of just under 300 °C. Acoustic emission data were collected using a microelectronics-based system to obviate the necessity for personnel to be present throughout the test. This equipment, which is a simplified version of acoustic emission instrumentation developed for the United States Federal Highway Administration, is capable of long-term continuous monitoring, followed by rapid data retrieval achieved from an LSI microcomputer memory.

Stainless steel piping of this type normally fails due to intergranular corrosion cracking, a failure mode common in power plant. Environmental conditions in the test loop (type 304 steel in mill condition, welded in a loop comprised of four lengths of 122 cm (48 in) long tube) were: temperature 274–280 °C, a dissolved oxygen level of 8·3 ppm and water conductivity lying between 0·1 and 0·5 mho cm^{-1}. Bending forces were applied hydraulically to the pipe until a bending stress of 196 MPa (28 500 psi) was achieved, representing 100% of the yield strength at test temperature (strain was measured at between 0·12 and 0·14). No cracking was detected under these conditions and so the maximum bending strength was increased to 268 MPa (38 800 psi), i.e. 136% of the yield strength at 288 °C. The measured strain then increased to just under 0·2. The test was carried out over a total of 54 days, during which time data were logged. While some background noise was identified, it was not difficult to separate the acoustic emission signals. No location device was used. The acoustic emission results were comparatively simple. For the first three days of the last 550 hours of tests the count rate was extremely low, less than a hundred events over 24 hours. After approximately five days the cumulative count rose to the 2 000 region and increased steadily over the next five hundred or so hours in a more or less linear manner. One day before the end of the test the count rate increased markedly from a few hundred per day to two or three thousand per day. When leakage occurred the count rate increased dramatically; six hours prior to pipe leakage the count rate increased by about a factor of 10 and, after leakage saturated the test equipment.

It was concluded from this investigation that the emission from stainless pipes subjected to stress–corrosion follows an expected pattern; no emission on the outset, constant accumulation during the apparent crack growth period and a distinct increase as the crack approaches full penetration of the pipe wall. The variation in count from one period to another implied that the emissions were produced by small crack increments. This conclusion was derived from the similarity of the pattern of emissions to that from slow crack growth due to the constant application of stress.

It is evident from Stahlkopf's work that acoustic emission can provide an indication of crack growth as, under stress–corrosion conditions, a crack approaches full penetration of the pipe wall, and that pressurised high-temperature water leakage from a through-wall crack is readily detectable. The overall test also demonstrated that the use of digital memory in an acoustic emission monitoring system is most valuable, providing a permanent record of the data in a way which can be rapidly evaluated when necessary.

It must be concluded that the use of acoustic emission to detect stress–corrosion in power plant pipe-work is a powerful technique which is simple and cheap to use. This is based not only on the results of Stahlkopf but also on my experience with other tests carried out under such conditions, which clearly indicate that stress–corrosion produces copious emissions.

In the detailed laboratory work of McIntyre and Green (private communication), an acoustic emission study under stress–corrosion conditions which is worthy of discussion in some detail, steel bars were cut into 13 mm CKS test pieces (longitudinal orientation) notched in the transfers rolling selection. Heat treatment at around 830 °C was followed by quenching in oil. After precracking by fatigue in air, test pieces were pulled either in a 3·5% salt solution in deionised water or in hydrogen gas at pressures in the region 0·25–1 bar (using a stainless steel chamber completely surrounding the test piece). Acoustic emission measurements were made with Dunegan Endevco equipment using transducers having a flat frequency response in the region of 0·1–0·3 MHz. Ring-down counts (monitored using a potential drop method) were not well correlated but total acoustic emission 'energy' was a good indicator of crack growth. The crack/length ratio and acoustic energy followed the same form as a function of time during stress–corrosion cracking of 897 M39 steel. It is interesting to note that McIntyre and Green also found that crack growth rate and acoustic energy were well correlated when plotted as a function of stress intensity during cracking of the same steel in the same series of tests. When energy was plotted as a function of cracked area, approximately linear relationships were obtained, the slope of which depended on the stress–corrosion crack path and the grain size of the steel. It was observed that higher levels of acoustic emission energy were recorded during intergranular stress–corrosion in the A151 430 and 817 M40 steel than during transgranular stress cracking in the M39 steel. In the case of the M40 steel higher levels of energy were recorded during crack growth in hydrogen where higher velocities were observed than for stress–corrosion cracking. In contrast, lower amounts of energy were released in hydrogen for the M39 steel (cracked trans-granularly) compared with those during stress–corrosion cracking. The same authors have also shown that structural steel cracking in a gaseous nitrate atmosphere also is readily detectable using acoustic emission.

It must be concluded therefore that the use of acoustic emission detection during stress–corrosion and other corrosive atmosphere related fractures is a most promising application.

4.5   Testing high-pressure underground pipelines

There is obviously a great need to provide a method for checking the integrity of underground high-pressure gas pipelines and several inves-tigations have been carried out, both in the field and in the laboratory to check the feasibility of using acoustic emission for this purpose. In the United States Parry carried out tests on a fifty year old pressurised pipeline belonging to the Philadelphia Electric Company (see Lehman 1974). It had been made using oxyacetylene welding. A transducer spacing of around 60 m (200 ft), in order to cover the line effectively, was achieved by using an acoustic probe mounted on the end of a long sharpened rod which could be pushed down through the earth to make contact with the pipeline. A 6600 ft (~2012 m) section of 12 inch (~30·5 cm) pressure distribution line was tested in two ways. In the first test, the line was pressurised to 120 psi (~827 kPa) using nitrogen; in the second, the line, which was buried about five feet beneath the surface, was loaded by stressing with a heavy vehicle moving up and down. The vehicle drove the whole length of the pipe with one set of wheels directly above the line.

In these tests 35 emission sources were located. Analysis of the results indicated that the locations did not threaten the integrity of the line. Following this test the line was excavated at locations where the six most prominent sources had been situated. The pipe welds were found to be coincident with each of the locations and the section of the pipe containing each weld was removed and capped for hydrostatic testing. Two control weld sections, not indicated by the acoustic emission analysis as containing sources, were also removed and subjected to hydrostatic testing. All eight sections were stressed cyclically to failure in the laboratory of the Philadelphia Electric Company. All six welds indicated by acoustic emission to contain discontinuities failed after a low number of cycles compared with the two control sections. This set of tests is some indication that acoustic emission is of value in detecting very bad defects buried in lines.

Laboratory work by Lumb and Hudgell (1976) and co-workers has also thrown light on this application of acoustic emission. The emission characteristics of stressed mild steels were studied, as were those of weld deposits laid down by the automatic submerged arc, manual rutile metal arc and manual CO_2 welding processes. The steels were those used in pipelines to API 5LX, grades X60 or X52. Tests were performed on compact K-fracture toughness specimens machined from line pipes without flattening, or from plate with the axis of the specimens in the axial direction of the pipe. The specimen thickness was 12·7 mm, the maximum possible from the material available. Lumb remarks that the total counts with these types of materials are generally low, consistent with other published information on similar steels. Most of the emission occurred shortly after maximum load with gross plastic flow thought to be the major factor contributing to the emission. The rate of emission did not increase with crack length or as failure approached. As one of the most common causes of failure of buried pipelines is interference by digging machines with consequent work-hardening of the material, the failure of a damaged pipe in a notched tensile specimen was investigated. In this case emission did increase immediately prior to failure but (according to the authors) not sufficiently long before failure to constitute a reliable damage indicator. The results on parent material did in fact indicate that for good quality material it would be very difficult to rely on acoustic emission for spotting damage between welds in a pipeline. However, if damage is found to occur in the weld the results are far more promising and the relative quietness of the parent material can be regarded as an advantage. The number of counts in the weld material was such that the material could not be regarded as acoustically quiet. In the case of submerged arc welds and manual CO_2

welds, significantly more emission occurred before general yield of the weld material than was noticed in the case of parent material. This is not true however for the manual rutile welds which are also acoustically rather quiet. It was further noted that stress-relieving (30 min at 650 °C) increased the amount of emission generally and in one case quite dramatically. It was evident from these tests that acoustic emission has good potential for monitoring crack growth in weld metals with certain alloy additions (e.g. niobium) during stress-relieving. Lumb also concludes that in mild steel structures acoustic emission has the potential 'to predict unequivocally and with reasonable anticipation, the imminence of failure by crack growth' in the heat-affected zone.

As a result of these laboratory and field tests it can be concluded that acoustic emission does have some value as a method of ascertaining the integrity of underground pipelines but that the results of acoustic emission investigations must be treated with care.

4.6   The detection of leaks using acoustic emission

There is no doubt that detection of leaks in water-cooling or gas-carrying systems is now an established technique. The work reported above in the United States on the detection of faults in gas lines amply confirms this. Nielsen, in a private communication, has reported success with the use of acoustic emission for the detection of through cracks in pipelines. In West Germany the complex cooling systems of nuclear reactors are regularly monitored using acoustic emission. To a purist this technique might not constitute 'acoustic emission' but there is little doubt that it is one of the most useful applications of the method and is likely to give a solution to frequently met industrial problems. The alternatives to acoustic emission for the detection of leaks have considerable difficulties associated with their application. For example the provision of flow meters, or gas or water pressure points is very expensive compared with the low cost of fitting a permanent or temporary acoustic emission transducer (the problems which can result from the high wiring cost for extended permanent acoustic emission installations can be overcome by using telemetry or by sending information down 240 V mains lines).

Leak detection is an extremely promising application for during regular operations; it has three advantages compared with the detection of 'normal' acoustic emissions:

(1) the signal from a leak is continuous not transient and can be detected at a low level;
(2) the background noise level is easily subtracted;
(3) a leak to atmosphere would always be expected to generate a strong acoustic signal.

One of the Westinghouse plants monitored by Gopal and his co-workers accidentally demonstrated the capabilities of acoustic monitoring in leak detection. During its cold hydro-tests a signficant leak of water started; the first indication to the plant operators was a series of very large acoustic emission bursts on all transducers.

4.7 The use of acoustic emission in tube making

A recent industrial application of acoustic emission has been in the manufacture of tubes used for the transport of high-pressure gases. These are about 1 m in diameter and are made of special low-sulphur, carbon–manganese grades of steel (denoted X60 or X70). During manufacture, each tube is work-hardened by pushing a mandrel into the tube, while cold, causing it to expand. With the cooperation of the makers (Italsider), workers at CSM Rome (M Mirabile, private communication) have placed transducers on the tubes to plot emission activity against mandrel pressure. They have been able to establish the degree of work-hardening of the tube material, which is not possible using other methods. This important application is something of a breakthrough for the technique as it is made 'on-line' during tube production.

The technique takes note of the common observation that, in materials such as X65 or X70, acoustic emission signals are generated intensely during the *initial* stages of plastic deformation. Consequently it is possible to define P_{AE} as the mandrel hydraulic pressure when the acoustic emission rate starts to rise rapidly and P_{max} as the maximum pressure observed. The results of a number of tests in one series were

Grade	P_{AE}	P_{max}	P_{AE}/P_{max}
X65	34·2	44·1	0·77
X70	55·8	66·1	0·84

(P in arbitrary units).

Mirabile has deduced that

$$n = [\lg(P_{max}/P_{AE})]/[\lg(\varepsilon_{max}/\varepsilon_{min})]$$

where n is the work-hardening factor of the material, ε_{min} is deformation at start of plastic deformation, giving $n = 0\cdot73$ for X65 and $n = 0\cdot11$ for X70 material. These values are in agreement with those of Brozzo (private communication) from static tests.

References

Birchon D, Dukes R and Taylor J 1974 *Inst. Mech. Engineers (London) Conf. Publicn* **8** p8

Gopal R, Smith J R and Rao G V 1976 *Inst. Mech. Engineers (London) Paper* C194/76

Holt J and Palmer I G 1974a *Inst. Mech. Engineers (London) Paper* C80/74

——1974b *Proc. Symp. on Acoustic Emission, Deutsche Gesellschaft für Metallkunde 1974*

Lehman E A 1974 *Pipeline Industry* June 1974

Lumb R F and Hudgell R J 1976 *8th World Conf. on Non-destructive Testing, Cannes 1976* Paper 3K17

de Michelis C and Mirabile M 1978 *Mater. Sci. Engng* **34** 213–26

Parry D L 1975 *ASTM, STP 571* p550 *et seq* (also *EXXON Nuclear Co. Report No.* XN-129)

Stahlkopf K and Green A 1976 *EPRI Journal* February 1976

Stahlkopf K, Hutton P H and Zebrowski E L 1976 *Inst. Mech. Engineers (London) Paper* C204/70

Watanabe T, Hashirizaki S and Arita H 1976 *Non-destr. Test. Int.* **9** 227–32

5 Monitoring the Welding Process

5.1 Introduction

Acoustic emission monitoring of the welding process promises to be one of the most important applications of the technique because it is thus possible to check weld integrity during the formation of the weld. Other non-destructive testing methods — especially ultrasonic and radiographic inspection — have proved to be very successful in detecting poor welds after they have been completed, but this means that it is possible to waste much prime material in the interval between welding and fault detection. 'In-weld' fault detection can also allow rapid feedback to control welding equipment parameters. When deep multi-pass welds are used it is important to detect faults in the early passes or else the whole weld will be ruined. The elegant and successful application of ultrasonics as detailed by Hetherington (private communication) and others can reduce these effects, but still leaves weld strength monitoring during the process as a desirable goal. It is possible to control welding by means of the easily measurable parameters of the process, such as welding current or time of contact, but these techniques could not be an adequate substitute for the comparatively direct measure of weld properties promised by acoustic emission technology.

Many welded structures are physically large — for example an offshore oil platform can have several *miles* of weld — and their inspection for weld defects is very expensive (5% of the total structure costs can be absorbed in inspection by ultrasonic and other means). Acoustic emission holds out the promise of pinpointing fault areas quickly by the use of fault localisation techniques; these areas can then be given a more detailed examination using radiography or ultrasonic methods. It can be seen why many workers have concentrated on this application; as welds constitute the most likely area of weakness in most large metal structures, their monitoring can be regarded as of paramount importance amongst all acoustic emission application areas currently being studied. This chapter will concentrate on applications of acoustic emission to the welding process itself, and to processes closely associated with welding, such as heat treatment, leaving general examination of welded structures to a later chapter. A detailed literature review of this subject has been given by Neumann *et al* (1975) who review the significant causes of weld defects (which are listed in table 5.1). Here 'working defects' are those due to 'operator' causes, and 'material' defects result from the base or weld materials. Further information on welding processes is outside the scope of this book and for this the reader is referred to more general texts on welding technology such as those by Nichols (1969) and Houldcroft (1977).

Table 5.1 Common defects in the welding of steels; the table shows only those kinds of defects the occurrence of which is likely to lead to sound emission.

Working defects	Materials defects	
	Ferritic steels	Austenitic steels
Slag inclusions	Cold cracks	Cracks in the
Pores	Hot cracks	embrittled structure
Bonding (joint) defects	Reheating cracks	Hot cracks
Copper inclusions	Terraced fractures	Reheating cracks
Tungsten inclusions		
Insufficient welding penetration		

The subject of weld monitoring will be dealt with in the following order:

(1) application to the monitoring of welds during formation;
(2) monitoring during stress relief to find cracks caused by reheating;
(3) surveillance immediately after welding to detect delayed crack formation.

5.2   Monitoring weld formation

The importance of in-weld monitoring to workers in acoustic emission is reflected by the early date of the first significant papers dealing with the subject; that of Hartbower *et al* (1971) in the ASTM publication STP505 and that by Jolly written as early as 1969. These publications stressed the main difficulty met in all welding applications of acoustic emission — that of severe interference from noise, generated by the process itself, or from electrical interference caused by, say, the high frequency 'starter' pulses used in many welding processes. All welding processes present workers in this field with a particular challenge, varying in severity with the welding process used. Three sources of process noise have been identified: (i) metal transfer, (ii) slag cracking and detachment, and (iii) noise caused by the operator. This process noise (normally causing reception of spurious signal amplitudes only below 3 mV) is less difficult to eliminate than electrical interference.

Nielsen (1976) in his excellent review of this application of acoustic emission, has classified welding processes by the severity of the expected interfering signals. Electro-slag welding is basically quiet. Electron-beam welding likewise is suitable for acoustic emission surveillance; it is free from material transfer and oxidation, and the process is carried out within a vacuum chamber, forming an efficient electrical shield for the acoustic emission equipment. This was confirmed by Kharchenko *et al* (1973) in the course of the monitoring of 1 mm thick sheets of niobium when welded to molybdenum. Electron-beam stainless steel weld monitoring has also been studied by Eisenblatter (1974) who could detect crack formation during welding, even when continuous emissions were taking place from a solidification process.

Spot welding is noisier than the above processes but as it is difficult to detect weak welds by monitoring welding parameters (such as current and electrode pressure) and because of its widespread use it is a particularly important area for the application of acoustic emission. Some success has been recorded in papers by Schwerk and Shearer (1970, 1973) by Steffens and Crostack (1976) and by Weaver and Knollman (1974) discussed in detail below. It has been reported that General Motors in the USA uses acoustic emission extensively to monitor the strength of spot welds using commercially available equipment.

Arc welding unfortunately is a copious producer of interference from high-frequency electrical signals, from material transfer and from oxide formation and cracking. However some workers have succeeded in overcoming these difficulties and have provided a basis for future applications; Romerell (1972) has described an application of acoustic emission to the automatic TIG welding of reactor fuel-pin end closures. Prine (1976) has also shown how, with care, acoustic emission can be applied to the important family of arc welding processes, and Wehrmeister (1977) has described its routine industrial application to monitor welds.

The application to each of these areas of welding technology will now be discussed in detail.

5.3 Spot welding

Spot (or resistance) welding is used extensively in the manufacture of cars and of light engineering assemblies. In more conventional applications of spot welding (motor-body manufacture for example) the weld failure rate is reduced by the use of redundant welds, in which case acoustic emission testing is not used. However, acoustic emission equipment is now widely used in the United States and West Germany for the testing of components requiring a very low in-service failure rate. As mentioned in the introduction to this chapter, a simple approach to the control of weld quality would involve the monitoring of weld current or the voltage drop across the weld during the current pulse. This method provides an indication of abnormal weld conditions but is not directly related to the processes which determine weld strength. In contrast, acoustic emission is related to metal flow and expulsion processes and can be related more easily to weld quality.

The Lockheed Corporation (see Weaver and Knollman 1974) has thoroughly investigated this application using equipment built by the Trodyne Corporation. This records the number of acoustic pulses generated by cracking after the weld is completed, as well as the number

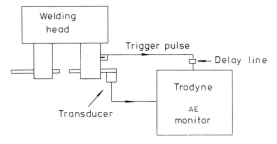

Figure 5.1 Apparatus used to investigate acoustic emission applied to spot welding.

of pulses produced by the normal flow of weld material. Also recorded are the number of high-amplitude pulses associated with expulsion of metal during the weld cycle. The Trodyne equipment uses a bandwidth of about 50 kHz centred on 225 kHz, and a transducer clamped to the weld electrode, as shown in figure 5.1. Silicone grease provides acoustic coupling between the electrode and transducer surfaces. A delayed-pulse generator was used to delay counting of acoustic pulses until after the start of the weld in order to exclude acoustic noise produced by electrode impact and cold flow of the weld material. Counting of pulses continues during the weld count interval up to one second after weld

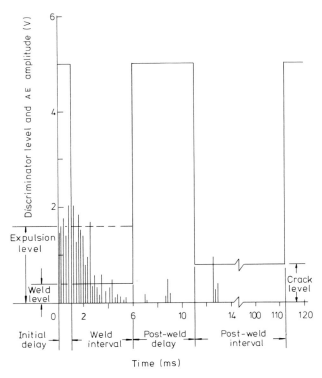

Figure 5.2 Timing sequence and discriminator levels for analysis of acoustic emission.

initiation. The strength of each weld was measured by stressing a pair of crossweld wires to failure.

Signals from spot welds between 500 μm nickel wires consist of random pulses as shown in figure 5.2. Emissions associated with the strong weld contain only few high-amplitude pulses; a medium strength weld shows a greater incidence of high-amplitude pulses while a weak weld has many high-amplitude pulses and a longer duration of emission than strong welds. In service, an on-line measure of weld strength can be derived from the number of high-amplitude pulses because weld strength tends to decrease abruptly as their number exceeds a predetermined maximum. Figure 5.3 shows a typical plot from an extensive series of tests on over 1 000 spot welds. In order to give a reliable rejection system which does not exclude too many satisfactory welds it is necessary

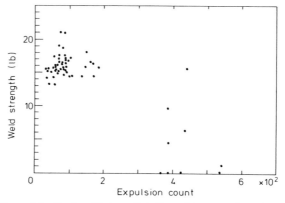

Figure 5.3   Spot weld strength indicated by acoustic emission.

to combine the criteria of a low expulsion count rate with a low total count. The Trodyne system described by Weaver and Knollman can be used in on-line analysis of spot welds. They point out that time-sharing of two or more monitor systems amongst many welding stations is feasible and does not require the use of special computing facilities. Schwerk and Schearer (1970, 1973) have described a similar, successful application of acoustic emission to the routine examination of spot welding. A more fundamental study of this application of acoustic emission was undertaken by Steffens and Crostack (1976) who identified

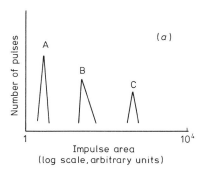

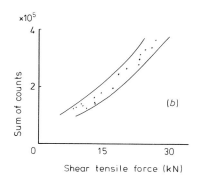

Figure 5.4 (*a*) Notional histogram of acoustic emission impulse area (spot welding): A, background noise; B, martensite formation; C, from magnetic field. (*b*) Correlation between weld strength and counts due to martensite formation (peak B in (*a*)).

the emission sources which can occur in spot welding and can be detected and identified using simple frequency and amplitude analysis. In particular they have shown that background noise can be isolated using frequency analysis. For example, magnetic field noises are associated with high frequencies and liquid metal sources give rise to low-frequency signals.

A more meaningful analysis can be made from the analysis of the size of the impulse area (see figure 5.4). The number of impulses from a spot weld is plotted against the impulse area. It can be seen from this figure that counts from the background, from the magnetic field and from cracking due to martensite can be differentiated. The figure also shows how the sum of the pulses formed during martensite growth is correlated with weld strength.

Recently Roeder, from the same school as Crostack, has developed an on-line frequency analysis approach to spot weld monitoring which shows promise of success. Impulses are placed in four categories — a good weld has a different set of characteristics from a poor one.

5.4 Arc welding

Arc welding is a widely applied method of joining metals. Account has to be taken of the differing types of arc weld used (generally TIG welding,

manual or submerged-arc welding). In general, workers concerned with
the quality monitoring of arc welds have concentrated on situations
where there is either a need for tight quality control of a specific weld
or where welds are carried out automatically without detailed operator
supervision.

This subject, one of the earliest studied in the field of acoustic emission
was reported in a classical paper by Jolly (1969). Figure 5.5 shows one
test result quoted by Jolly in which an acceptable weld gave only 77
emissions over 7 passes; rogue metals in subsequent welds were added
to cause hot cracking during solidification of the weld bead. The graphs,
of which figure 5.5 is typical, show that the addition of tantalum and
titanium is followed by a considerable increase in acoustic emission.
Titanium, when added on the third pass of one good quality weld, caused

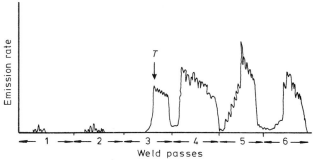

Figure 5.5   Effect of poor weld on acoustic emission rate. Titanium added
at T.

the number of emissions to rise to a total of $1 \cdot 8 \times 10^3$ from around a
hundred. Steffens and Crostack (1976) have also correlated total counts
with the number of metallographically determined cracks in a sub-arc
weld.

Recent work by Baruch *et al* (1976) has shown that acoustic emission
is very effective in detecting cracks in arc welds. Experiments were
carried out using gas tungsten-arc welding on 2 mm thick sheets of
austenitic stainless steel. Cracks were produced at predetermined locations by diluting the base metal with an impurity (copper) by drilling
holes along the weld path and inserting copper plugs. Argon was used
to shield the arc, preventing interference from slag cracking; welds were
made with an automatic head travelling at a fixed speed. A novel feature

of this work is the use of a paired ferro-electric transducer instead of a PZ crystal. In these results, peaks in the emission rate corresponded to weld cracking, but Baruch found that the peak height was not easily related to crack length. These results were not so satisfactory with type 302 stainless steel because of background noise caused by martensite formation under welding stressing. Unfortunately, Baruch did not give details for all the alloys he tested.

Prine (1976) has studied the application of acoustic emission to the monitoring of nuclear power plant piping welds. This is an example of the use of acoustic emission to monitor welds under 'high quality' conditions, when failure in service must be avoided at all costs. Standard commercial automatic welding equipment was used, with the inclusion of a PZ transducer on the weld head. A series of test welds was monitored, some of which were intentionally flawed. The bead materials used were A106 carbon steel and A312 T304 stainless steel.

The four welding methods chosen for these tests range from acoustically quiet TIG to the noisy submerged-arc and manual stick welds. All welds were multi-pass, averaging five passes, and 100 welds were monitored in all. All the tests were performed in a nuclear-certified welding shop using certified materials, and the finished welds were radiographed according to ASME nuclear standards. In addition to the calibration tests, two weeks of normal production nuclear welding on similar piping were monitored in two separate nuclear pipe fabrication shops. The calibration samples were available for additional NDT or destructive testing and standard ASME nuclear code results on the production welds were made available.

Signals within a 100–400 kHz band were recorded and an alarm signal registered when any event fell within an amplitude window (set during the course of the tests). The alarm did not go off when the pipe was hit or ground, but only when faults were induced in the calibration test welds by contamination with copper. Frequently, the cracking induced by the copper not only produced easily detectable acoustic emission in the pass in which it was introduced but continued to produce activity throughout several subsequent passes. An example of this occurs in a 61 cm (2 ft) outside-diameter carbon steel pipe. Copper was introduced on a TIG root pass and produced some cracking as well as acoustic emission activity. Then the automatic submerged-arc was used to complete the weld, and the copper produced cracks and acoustic emission alarms during three of the five passes needed to finish the weld.

The ability of acoustic emission to detect cracks which are difficult, if not impossible, to detect by conventional NDT methods is illustrated in results obtained on a MIG weld on a 30·5 cm (1 ft) stainless steel test weld. Copper was introduced during welding, and acoustic emission was produced. ASME code radiography performed on the weld failed to detect the crack. Careful laboratory radiography performed later still detected no crack (the quiet nature of 304 stainless steel was noted in this test). Copper was introduced in the later portion of the weld and a single acoustic emission alarm occurred about 10 s after the introduction of the copper. Metallography confirmed that the crack detected was of an interpass type and in a plane parallel to the weld surface, making it difficult to detect radiographically.

In addition to the calibration welds, two weeks of real nuclear pipe production welding was monitored by Prine in two separate pipe fabrication shops. A total of 11 acoustic emission indications were produced, all of which correlated with either visual, dye penetrant, or ASME code radiographic results. Six indications were obtained during a series of TIG repairs on a heavy-walled stainless steel section of primary loop piping. The cracks were confirmed by dye penetrant and were eventually ground out and successfully repaired. One indication was produced by small radial cracks found in a TIG root weld caused by improper arc break-off. The cracks were visually confirmed and repaired on the spot by the welder. The remaining four acoustic emission indications were all due to slag inclusions which were not rejected under our ASME code. The slag inclusion indications could probably have been eliminated by a slight reduction in the gain of the acoustic emission monitor.

Acoustic emission has been adopted by the Babcock and Wilcox Company to monitor multi-pass submerged-arc welds, described in a paper by Wehrmeister (1977). In this application, welds in 15 cm (6 in) thick carbon steel plates were monitored with novel acoustic emission sensors that travel with a welding electrode. The detection system provides flaw location capability for immediate weld repair. In many respects this work is similar to that described above but the use of a novel transducer is most interesting. It combines the techniques of acoustic emission with electro-acoustic (EMA) pick-up of the emitted signals. Conventional liquid coupling of transducer to workpiece was found to be impracticable because of difficulties associated with controlling the flow of liquids when near the weld zone.

In contrast the EMA sensor does not require a liquid coupling medium, is self-supporting and noncontact. Babcock and Wilcox found that the probe was easily adapted to move along welds even in an industrial welding plant. The stand-off distance of 0·05 cm (0·02 in) from the material was maintained by a three-leg support. An air jet was directed at the face of the transducer for cooling purposes and to remove any welding flux which might catch and drag beneath the sensor gap. The EMA transducer is, however, about 40 dB less sensitive than the equivalent PZ sensor, and is sensitive to welding currents and to airborne electrical interference. A microprocessor was used in this system to allow rejection of some signals. Signal validity is based on checking the location of sources (which must lie within the weld zone).

This work indicated that cracks and slag entrapments in multi-pass submerged-arc welds will be detected with a high level of confidence by acoustic emission. Porosity and lack of fusion may be detected if associated with cracking. However the use of special signal processing and source location techniques is necessary for the use of acoustic emission in a shop environment. Movable EMA sensors are practical, and necessary for thick-section circumferential welds if wrapping cables around the component is to be avoided. The use of a programmable microprocessor is considered a distinct advantage for system adaptation to the variety of weld configurations found in normal practice.

The principle of an EMA transducer (see Whittington 1978) is shown in the diagram, figure 5.6. The essential features are a powerful magnet and a small flat coil of wire in close proximity to the surface of the test piece. A pulse of alternating current is passed through the coil, inducing a similar current in the surface of the test piece (which must, of course, be electrically conducting). Since a current-carrying conductor will experience a force when placed in a magnetic field, alternating stresses will appear in the test piece at the same frequency as the applied current pulse. Thus the test piece will vibrate and an acoustic pulse will be transmitted through it. The same system may be used for receiving. When a conductor moves in a magnetic field, an EMF will be generated in it. Thus vibration of the surface due to the arrival of a pulse of ultrasound under the transducer will set up an alternating current, which will be picked up in the coil. The magnitude of these effects will depend on the strength of the magnetic field, the electrical conductivity of the test specimen, the current in the coil and its distance from the surface.

Roughly speaking, when using a field strength of around 0·5 T, and with the coil 1 mm from the surface, the received signal strength will be about 40 dB down on that of a conventional PZ system using equivalent transmitter power. This lack of signal strength may be overcome, however, by using a sufficiently powerful transmitted current pulse, so that the signal-to-noise ratio obtainable with EMA can be as good as, or better, than that from conventional ultrasonic systems.

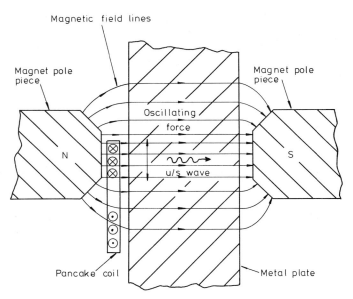

Figure 5.6   Principles of EMA generation.

Rommerell (1973) has also studied the use of acoustic emission to monitor the automatic production of small TIG welds, in the production of nuclear fuel-pin tubes; here quality control is evidently of the greatest importance. He concluded that 'Acoustic emission monitoring augments other non-destructive testing methods and is sometimes applicable when other tests cannot be applied. This is, in part, due to the high sensitivity of acoustic emission monitoring. Acoustic emission monitoring is only sensitive to active flaw growth, however, and will not detect a flaw in equilibrium.'

5.5 Electron-beam welding

During electron-beam welding, various acoustic sources occur whose impulses can be differentiated by their amplitudes (Rommerell 1973); a very strong signal is encountered in electron-beam welding, caused by the evaporation of the material underneath the beam, owing to phenomena within the weld pool. Signals from electron-beam welding originate in material movements within the weld pool and, hence, can be utilised for process monitoring or for parameter control. Since the kinetics of weld pool movements directly affects the formation of the weld seam, there is also a direct relationship between acoustic emission and the penetration depth of the electron beam into the work piece. Individual welding parameters such as beam velocity, the beam focus or the beam current can be directly related to acoustic emission with respect to this effect on the penetration depth of the beam (Rommerell 1973). This phenomenon can be utilised in automatic welding, e.g. in the control of feed rate or in automatic detection of the gap between the two work pieces being welded (which, as a rule, is very small in electron-beam welding). It is also possible to control the focus of the beam (focusing is particularly difficult in the case of large beam currents). The correct application of electron-beam welding depends critically on correct focusing and as the position of the focus alters with fluctuation of the working distance between the cathode and the work piece (for example, because of surface irregularities), the focus must be closely controlled. Acoustic emission makes this feasible.

5.6 The use of acoustic emission to monitor stress relief

Laboratory investigations have attempted with some success to follow crack formation in the heat-affected zones of welds during stress relief treatment. Despite the advantage of being able to accurately determine the time of onset of cracks, it is a difficult technique to apply in practice owing to the necessity to operate at high temperature. Industrial application of the technique is not so widespread as is the case for weld monitoring. However, Glover *et al* (1974) have carried out some work on a mild steel header retracted from a power station, held at a temperature of 565 °C and under a pressure of 0·25 GPa (2·5 kbar). In the test furnace, stainless steel waveguides were welded to the header to give acoustic emission signals during pressurising and depressurising. Crack formation and growth were easily detected using this method of 'tapping' into the equipment. Jax (1973) has also reported some success in the evaluation of cracks formed during stress relief.

5.7   Monitoring delayed cracking in welds

This subject will be treated more thoroughly in a later chapter (concerning the monitoring of offshore structures). It is interesting to note that Birchon (1976) has drawn attention to the usefulness of acoustic emission in detecting welding cracks several days before they are large enough to be found by ultrasonic inspection. He has found crack initiation in the heat-affected zone within one hour of the completion of welding.

The possibility of early detection of cracks in welded structures has been studied by Lumb and Hudgell (1976) who determined the emission characteristics of some low-alloy steels with weld deposits led by submerged-arc, manual rutile metal arc and the manual CO_2 processes. In their work, which was specifically concerned with pipeline steels (grades X60 and X52), signals were analysed by ring-down counting using standard Dunegan Endevco signal channel equipment. Their most important conclusion was that general corrosion or mechanical damage could be detected using acoustic emission. In some cases failure of buried pipelines is caused by digging machines, and the authors stated that they do not anticipate acoustic emission being of value in detecting failures from this cause. They did not show a correlation between cracks in 'clean' plate (i.e. low inclusion density) but they did find emission associated with gross plastic flow. Dirty steel plate gives rise to copious emission prior to general yield.

Overall, this paper has laid a good foundation for the realistic application of acoustic emission to the detection of faulty welds in structural steels and the results can be regarded, on the whole, as encouraging for the usefulness of the technique. Their important conclusions can be summarised as follows:

(1)   the overall level of acoustic emission in clean mild steels, which comply with the API 5LX specification, is generally low, but is somewhat higher in weld deposits;

(2) there is little emission during crack growth and no general increase in the rate of emission with increasing crack length, both in clean plate (API 5LX) and in as-laid deposits;
(3) in both plate and welds, much emission appears to be associated with gross plastic flow;
(4) neither clean plate nor manual rutile metal arc weld deposits give sufficient emission prior to general yield to suggest an attractive basis for defect detection in proof tests, but there appears to be sufficient emission for a reasonable basis for defect detection in both manual CO_2 and submerged-arc welds;
(5) dirty steel plate gives rise to copious emission, mostly prior to general yield;
(6) copious emission occurs during crack growth in some submerged-arc weld deposits embrittled by stress relieving at 650 °C and also from cracks growing in heat-affected zones;
(7) cracks growing in heat-affected zones are accompanied by an increasing rate of emission as fracture approaches.

5.8 Summary

Almost all structures, vessels, goods, pipelines and vehicles made of steel involve welds. These are usually the weak points in their construction and it is rare for failure to occur in the parent metal but not unusual for cracks to start and grow in a weld or its heat-affected zone.

From its earliest days, acoustic emission has been applied to the monitoring of welding, despite the presence of considerable background noise in most welding processes. The electro-slag and electron-beam methods are both acoustically quiet and are used in 'high-integrity situations'. These factors make them good subjects for acoustic emission monitoring and many papers have described its successful application.

Spot (resistance) welding is used extensively in the manufacture of automobiles and light engineering assemblies. Acoustic emission can be successfully used to monitor the production of spot welds where there is a need for tight quality control in the assembly of high-reliability components. The high background noise, both mechanical and electrical, associated with spot welding can be eliminated by timing the period when emissions are monitored. Acoustic emission monitoring of more conventional instances of spot welding, such as in motor-body manufacture, has also been successful in some cases, although this application has been slow to gain ground.

The arc welding of nuclear equipment is successfully monitored by acoustic emission and in one instance special equipment has been developed to monitor multi-pass submerged-arc welds used in pressure vessel manufacture.

To sum up finally, acoustic emission can be readily used in the less noisy welding processes such as electron-beam welding. The other, more common welding methods can be monitored in this way, but care has to be taken to eliminate effects due to background noise. When a really high-integrity weld is required, as for example in nuclear applications, acoustic emission is now accepted as a valuable aid to ensure good welds.

References

Baruch J, Yaron S and Golan S 1976 *Proc. 8th World Conf. on Non-destructive Testing, Cannes 1976*

Birchon D 1976 *Acoustic Emission* ed R W Nichols (Barking, Essex: Applied Science)

Crostack H A 1974 *Proc. 3rd Mtg of EWGAE, Ispra 1974*

Eisenblatter J 1974 *Industriell Anwending in Schallemissions Analyse* (Oberursel/Taunus: Deutsche Gesellschaft für Metallkunde) pp 222–53

Glover A G, Holt J and Williams J A 1974 *Schallemission Anwending bei der Unternchung Prüfung* (Oberursel/Taunus: Deutsche Gesellschaft für Metallkunde) pp 118–32

Hartbower C E, Reuter W G, Marais C F and Crummins P P 1972 *ASTM, STP 505* pp 187–221

Houldcroft P T 1977 *Welding Process Technology* (Cambridge University Press)

Jax P 1973 *2nd Int. Conf. Structural Mechanics in Reactor Technology, Berlin 1973,* Section G 614

Jolly W D 1969 *Weld. J.* **48** 21–7

Kharchenko G K, Zaderii B A and Kotenko S S 1973 *Autom. Weld.* **26** 70–1

Lumb R F and Hudgell R J 1976 *Proc. 8th World Conf. on Non-destructive Testing, Cannes 1976* Paper 3K17

Neumann E, Neumann V, Nabel E and Eisenblatter J 1975 *Lecture to Deutsche Gesellschaft für Zerstörungsfreie Prüfuerfahren, Berlin 6–7 May 1975*

Nichols R W 1969 *Weld. Metal Fabric.* **19** 344–51

Neilsen A 1976 *Acoustic Emission* ed R W Nichols (Barking, Essex: Applied Science)

Prine D W 1976 *Non-destr. Test. Int.* **9** 281

Rommerell D M 1972 *Mater. Eval.* **30** 254–8

—— 1973 *Weld J., Weld. Res. Suppl.* 81

Schwerk E B and Schearer G D 1970 *Weld. Des. Fabric.* **40** 35–6

—— 1973 *Non-destr. Test. Int.* **6** 29–33

Steffens H D and Crostack H A 1976 *Arch. EisenhuttWes.* **47** 12

Weaver J L and Knollman G C 1974 *Lockheed (Palo Alto Research Laboratory) Report* LMSC-D 358 302

Wehrmeister J 1977 *Mater. Eval.* **35** 45–7

Whittington K R 1978 *Phys. Technol.* **9** 62

6 Offshore Applications

6.1 Introduction

Offshore structures are used primarily to reclaim gas or oil from the sea bottom, and are normally of tubular steel or of reinforced concrete construction. Their size varies from a few metres high in the case of near-shore applications to a height of several hundred metres in structures in the upper regions of the North Sea. This chapter is little concerned with concrete platforms as the structural integrity inspection of these is still at a rudimentary stage, partially because of the long and successful history of steel structures in the southern North Sea and in the Gulf of Mexico, and the increasingly wide use of steel jackets. The inspection of steel offshore structures in the arduous conditions of the North Sea presents a considerable challenge for the acoustic emission test. There is arguably no other technique capable of monitoring these structures, although one competing method (some regard it as complementary to acoustic emission) will be discussed during the course of this chapter.

Most steel offshore structures are of the steel jacket type, illustrated in figure 6.1. Newer forms of construction using a large buoy tethered to the seabed and submarine equipment (in which all operations are carried out in a structure on the sea floor) are not as yet available. Some hundreds of jacket structures have been installed around the world. The commonest welded connection in the steel jacket type of structure is a simple joint in which the tubular members are welded together, with all the load being transferred from one branch to another by the cord, usually without any help from stiffening rings. Offshore conditions give rise to cyclic loads varying with wind and wave configuration. The peak strain may be from about 2·5 to over 25 times the nominal cyclic strain. This leads to the possibility of failure by fatigue in certain well defined critical areas of structures. These should be monitored using acoustic emission methods. Acoustic emission can also be of value in monitoring the integrity of the jacket's superstructure and its equipment, especially cranes.

This is where acoustic emission comes into its own, for it is possible to continuously monitor structures round the clock and to maintain monitoring even when weather conditions are bad. Inspection by divers is a hazardous business and they have to concentrate to maintain their life-support systems during inspection. While a high degree of concentration on the task in hand is obtained from divers with good inspection results, it is true to say that visual inspection of an offshore structure by divers is never carried out under the ideal conditions! The NDT techniques used are limited and are essentially 'on land' methods. The development of better and more complex methods to ensure the structural integrity of oil production platforms is necessary to meet the United Kingdom offshore installation regulations of 1974. These require that installations shall only be established on the United Kingdom continental shelf which have a valid certificate of fitness, requiring that areas of stress concentration and welded joints are subjected to 100% non-destructive testing. Structures would only qualify for recertification after a 100% inspection of all major load areas every five years. For platforms already in position an annual survey of 10% of the total number of joints supplements the five year survey before recertification is granted. Acoustic emission can play a major role in offshore safety because of its ability to pinpoint areas of concern which need to be examined more closely by divers.

The author is fortunate in having access to the work of Rogers, Webborn and their colleagues of the Unit Inspection Company, and he is grateful to them for permission to reproduce many of the results from experiments on a specially built underwater fatigue test rig using a tubular T-junction subjected to fatigue forces produced by surface waves. Also reported are background noise measurements from a jacket platform in the Danish gas-producing field and from an oil-producing platform in the Viking Field, both in the North Sea.

6.2 Inspection methods

The four alternative approaches to offshore structure inspection are: (i) inspection by divers, (ii) inspection from a submersible, (iii) acoustic emission, and (iv) the use of vibration monitoring. These will be discussed in turn.

Divers will always be required for complex mechanical work and

detailed inspection of welds, particularly in those regions shielded by
ancillary structures (for example the piles and pile guides encasing the
main legs). However, divers can only operate for a very short period
of time. At the moment they use simple magnetic testing which requires
that marine life forms be removed from the structure before testing,
and of course they can only operate under very favourable weather
conditions. Even the most skilled and dedicated divers have of necessity
to be very concerned about the maintenance of their life-support system
and consequently they cannot give as much attention to inspection as,
say, a land-based inspector. 100% inspection using divers is nearly
impossible in the view of many operators.

Manned and unmanned submersibles such as the Vickers Pisces and
Perry class vessels, and the Cetus vessels are being used increasingly for
inspecting deep-sea structures. While electronic and hydraulic attach-
ments make them more suitable than divers for many tasks they are
unlikely to completely supersede divers. However the use of the dry
'shirt sleeves' environment of a submersible certainly makes their use
very attractive. However, it is thought unlikely that submersibles will
be able to carry out the complete inspection required under the regulations without the use of more sophisticated examination methods than
are possible at the moment.

Turning to vibration monitoring, this method has been pioneered by
two groups in the United Kingdom. Little has been published about the
technique and it is thought unlikely to be able to provide the very
detailed non-destructive testing of critical areas that the regulations will
call for. In fact, a structure must be near collapse before the existence
of a dangerous situation would be drawn to the operator's attention.

6.3   Progress with acoustic emission monitoring

The use of acoustic emission is now steadily becoming accepted as a
valuable back-up to diver and submersible inspection. Its great advantage
is that a comparatively small number of sensors, spaced a metre or so
apart around a critical area, can keep a 24-hour, all-weather watch on
a structure. Experience has already shown that this method can pinpoint
areas which deserve examination by divers who can then go down during
a patch of clear weather to carry out a magnetic, visual or other inspection
of an area pinpointed to within less than a foot at the worst. This has

been demonstrated by the joint work of the Unit Inspection Company and the Danish Welding Institute on a structure in the Danish field, described below. The type of structure to be examined is shown in schematic form in figure 6.1 where a particular node subject to acoustic emission examination is also shown in detail.

The Unit Inspection Company, have studied the feasibility of monitoring using acoustic emission for offshore structures. Extensive laboratory and field work has been conducted at their own laboratory, the laboratories of the British Steel Corporation, at Imperial College, at

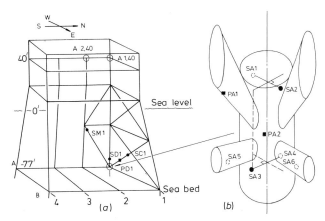

Figure 6.1 Schematics of (*a*) platform BD and (*b*) node A1−77' showing the sensor (S) and pulser (P) positions.

the National Engineering Laboratory and offshore, in collaboration with the National Maritime Laboratory at Totland Bay, Isle of Wight and in the southern North Sea in collaboration with the Continental Oil Corporation (USA). The material has almost invariably been of BS4360 50D and 43C steels in the as-welded condition. These extensive tests have shown that:

(1) both crack opening and crack closure fatigue, and coercion fatigue in structural steels in the BS4360 range produce emissions of detectable amplitude;

(2) problems with high levels of background water-borne noise, the coupling of this noise into the structure and mechanical noise

originating from topside machinery, cranes, gas valves, generators and pumps, and wave impact with the structure, can be overcome by careful choice of radio frequency, transducer design and mounting techniques;

(3) electromagnetic interference and crosstalk between signal leads can be overcome using active sensors, adequate screening and the avoidance of cable interconnections;

(4) it is possible for all electrical components and cabling to be safe in 'division 1' hazard areas and also to meet the requirements of long-term system durability — this includes resilience to water ingress corrosion, wave damage in the splash zone and electro-chemical compatibility with the structure;

(5) a wide variety of node geometries and sizes can be monitored using a diamond configuration of sensors without the necessity for guard transducers (it is possible that guard transducers might have to be used within the splash zone) — this is a most important item as it means that coverage of 48 critical nodes in a major structure can be achieved by using only 240 sensors.

I will now discuss the work carried out by Rogers, Webborn and their colleagues (see Rogers 1979) starting with the tests carried out at the National Engineering Laboratory on a large tubular T-joint fatigued in air.

6.3.1 NEL tests

For these tests BS4360 43C material was used and the scale of the test was tubes of 114 mm outside diameter, 5·5 mm thick, with a brace tube of 60 mm outside diameter, 4·9 mm thick. Specimens were fabricated by welding a 200 mm brace tube to a cord tube 740 mm long to form a symmetrical T configuration. A 100 kN servohydraulic unit under close loop control available at the National Engineering Laboratory, East Kilbride was used for the tests. The welds were formed using manual metal arc processes, without pre- or post-welding heat treatment. Constant frequency, random amplitude loading was used. Emissions were only counted provided they satisfied the following criteria: (i) the emission source lay within the weld area as determined by the Dunegan Endevco 1032 location mechanism used, and (ii) the emissions occurred at peak load. A typical result showed that the number of emissions increased markedly at 68% of the lifetime.

6.3.2 Controlled sea environment tests

Turning to the fatigue tests on T-joint specimens placed in the English Channel, fatigue failure has been produced in the same type of tubular joints as studied at the National Engineering Laboratory but in a real sea environment. A fatigue rig has been built and installed on the seabed 100 m off Totland Bay pier in the Isle of Wight. The rig was 4·7 m below mean high water, at a sight which was chosen to be close to the British National Maritime Institute environmental monitoring facility at the end of the pier, which continuously monitored wave and tide height, wind speed and wind direction. The top of the test rig was 3·0 m above the seabed; a loading disc 1·5 m in diameter was needed to apply force to the specimen at wave frequency. The rig was situated such that the paddle faced the predominant current and wavefront direction. The rig was designed to fail specimens in approximately four weeks, the time to failure, however, ultimately being dependent on the sea state. A major feature of the design was ease of specimen changeover, since this operation required the use of divers working in poor visibility. Other design features included: (i) extensive precautions against water ingress to the transducers, preamplifiers and signal cables, and (ii) acoustic isolation of the specimen from mechanical noise generated by the magnetic interference.

Dunegan Endevco S140 sensors and DE 3000 series analytical instrumentation were used to locate and characterise the acoustic emissions. The loading on the brace was monitored using Ailtech weldable strain gauges which provided the input to a voltage-controlled gate in the detection equipment. This allowed segregation of the crack closure and crack opening emissions, the instrument counters being engaged only at predetermined times in the loading cycle. Pressurised cabling was used to relay the acoustic emission signals to the instrumentation facility on the pier. The measurement equipment was essentially the same as that used for the in-air tests carried out at NEL, except that data were dumped periodically on to paper tape so that the instruments could be left unmanned, monitoring acoustic emissions continuously for periods up to two weeks.

Six samples were studied over a period of twenty months from January 1976 to August 1977. In order to register when a through-crack had developed, the sample was sealed and pressurised, a pressure transducer being used to record that point in time when the system was depressurised as a result of the fatigue crack becoming a 'through-the-thickness' crack.

Despite the precautions taken to isolate acoustically the specimen from the test rig, sources of extraneous noise were found to exist. These have been found to be mainly associated with the sea state, increasing with wave height and also with a low tide condition. The efficiency of the test rig was greatest at low tide when maximum load was exerted on the disc. Thus the sea state which produced the greatest fatigue, also produced the worst case of background noise. It was not possible to use a high threshold level setting to discriminate against spurious noise because the emission amplitude from the growing fatigue crack would not have been detected above the very high threshold needed to eliminate this background noise. It was found necessary to have a transducer arrangement which was asymmetrical in order to distinguish between the effects of wave-borne noise (see figure 6.2).

The Totland Bay fatigue testing of tubular T-specimens of BS 4360 43C steel have shown that plastic zone formation and crack propagation in the weld is accompanied by relatively prolific acoustic emission.

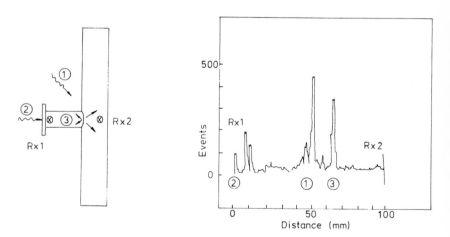

Figure 6.2 Typical result of linear location between sensors Rx1 and Rx2 for a cracked sample. The transducer positions used for the location measurements were chosen such that the spurious noise due to electromagnetic interference (1) and loading of the machine fixtures (2) was eliminated from the weld area, thus permitting the crack emissions (3) to be resolved.

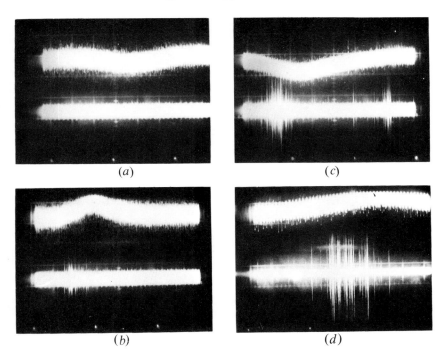

Figure 6.3 Strain gauge output (upper trace) and corresponding acoustic emission (lower trace) for: (*a*) new (uncracked) specimens under compressive loading — no emissions; (*b*) new (uncracked) specimen under tensile loading — emissions observed at rising load are spurious, probably originating from the loading train; (*c*) cracked specimen during compressive loading cycle — large number of emissions; (*d*) cracked specimen during tensile loading cycle — large burst of emission near peak load (notice reduced gain).

Emissions were observed during the tensile and compressive parts of the loading cycle, indicating that frictional rubbing of the crack faces, and grinding of corrosion products in the crack can be used as a measure of crack size. This can be seen from figure 6.3.

6.3.3 Tests on jacket structure ('Viking' experiment)

A proving ground for component durability, system design and method of installation was provided by the installation on the Viking Field structure. The aims of this experiment were:

(1) to gain experience in designing to meet intrinsic safety requirements in 'division 1' areas;

(2) to evaluate the precautions taken against: (i) water ingress; (ii) electromagnetic interference from power cables and radio transmissions, (iii) electrochemical reaction with the jacket, and to evaluate cable damage in the splash zone and the adequacy of the cable clamps;

(3) to evaluate the sensor design, in particular, the necessity for environmental acoustic shielding and additional line drive amplification;

(4) to recommend improved transducer configurations and analytical procedures in the light of measurements made;

and above all,

(5) to validate data in the presence of high water-borne background noise.

In general the number of background noise sources increases with decreasing frequency, and the energy spectrum of most individual noise sources also increases with decreasing frequency. By contrast the energy spectrum of legitimate acoustic emission sources is essentially flat up to very high frequencies, so the first basic approach to overcoming noise is to raise the operating frequency. With increasing frequency, however, the structural attenuation also increases, so a higher density of sensors is required to ensure that a sufficient number of sensors lie within detection range of the source. Thus there is a basic trade-off between operational noise problems and transducer spacing (and hence system cost), with operating frequency as the variable parameter.

At the operating frequency of the system, 60–180 kHz, it has been found that mechanical structure-borne noise was undetectable below sea level and just detectable at cellar deck level when a crane was operating. Damping of surface elastic waves by the seawater was high, around 20 dB m^{-1} in the near field (< 1.5 m) and 5 dB m^{-1} in the far field regions, but coupling of sea-borne noise to the jacket legs and braces was low (20 dB in the worst case of normal incidence). The extensive marine growth generated no detectable noise. The extent of wave distortion on passing under and across acute and obtuse fillet welds was markedly different, the attenuation varying between 2 and 20 dB, depending on the type of weld. Good correlation was obtained between the measured transit times of pulsed sound in the steel taking the shortest transmission path from pulser to sensor, and the timing on either the leading edge or the most significant peak. Using a fixed 1 V threshold resulted in greatest uncertainty because of differences in the leading edge and peak wave velocity, measured in steel below water to be 3·76 and 2·87 m s^{-1} respectively.

The underwater sensors were swamped by the sea-borne noise generated by work-boats (100 events/s at the sensor). Splash zone noise in the most severe sea states was appreciably lower (1–10 events/s). As the sensors were not acoustically shielded from the water, this noise resulted in frequent acceptance of Δt sets by the DE 1032 system as being valid data. *However, these events can readily be rejected by 'velocity validation'*, i.e. computing the event wave velocity at each sensor and rejecting those sources which did not have a consistent wave velocity within stipulated minimum and maximum limits.

It has become apparent from this work that structural integrity monitoring offshore is viable. The events showed a definite clustering at the 7 o'clock position on the left-hand diagonal branch weld; however, recent diver inspection of this area revealed no visible sign of cracking. Undercutting at the 12 o'clock position on the horizontal branch immediately below the suspect area had been detected the previous summer and ground out. We regard the absence of any visible sign of cracking as not contradicting the observed acoustic emissions which were of low intensity and indicative of a minor defect at worst. Data are, however, urgently needed on 'real' crack emission amplitudes under practical loading conditions. Guard transducers on the branches are not required, except possibly in the splash zone, four sensors and one pulser being adequate for monitoring nodes with four to six branches. The intertransducer distance should not exceed three metres. The sensors should be of the active type, the signal cables being superscreened from the sensors to the instrumentation. Breaking of cable continuity should be avoided and the pulser supply cables should be kept separate from the sensor cables to avoid crosstalk. Acoustic shielding of the sensors from the sea-borne noise is recommended in order to improve the availability of the system computer for real emission source identification.

The principal advantages of acoustic emission over other forms of non-destructive testing are the wide area surveyed and the real time nature of the method. It detects only those defects made active by the working loads imposed on the structure and, therefore, lends itself readily to remote, continuous monitoring.

6.4   Summary

The rewards from using acoustic emission to monitor offshore structures are considerable, but unfortunately the difficulties are also very great. Acoustic emission has the capability of providing all-weather monitoring services for structures which, by their very nature, cannot be allowed to fail without adequate warning to operators. Not only is the expense incurred by the lack of serviceability of offshore structures very great, but lives can easily be put at risk by undetected structural defects. The use of divers as the sole method of monitoring these structures will probably not continue indefinitely and there is pressure on operators to find alternative ways of checking the integrity of offshore platforms. The techniques developed using vibration measurements have not yet proved to be as effective as acoustic emission.

In contrast, tests carried out on large-scale laboratory structures subjected to fatigue stresses and on a scale model placed in realistic offshore conditions have shown that acoustic emission is effective in detecting the type of crack growth experienced in offshore structures. Tests carried out on the structures themselves have confirmed this view and have been extended to show the usefulness of the method for checking defects in rig superstructures.

Reference

Rogers L M 1979 *Proc. Int. Conf. Machine-aided Image Analysis, Oxford 1978: Inst. Phys. Conf. Ser. No.* 44 pp 151–167 (see also *Unit Inspection Company Technical Report No.* 1048)

7 The Application of Acoustic Emission to Aircraft Structures

7.1 Conventional airframes

The monitoring of aircraft structures constitutes a major challenge to acoustic emission since the benefits of the successful use of the technique would be considerable. Many examples are known where cracks have occurred in structures which have been located in parts of airplanes beyond the scope of normal, visual inspection. In other cases cracks have been undetected despite regular inspection using eddy current and other NDT techniques. In many aircraft small cracks can grow rapidly in hidden locations, under fastener heads, between lap joints etc, as well as in completely inaccessible parts, such as rear bulkheads. Acoustic emission is an attractive technique for monitoring aircraft structures because of the potential improvements in structure safety and potential cost savings that can be achieved by locating flaws quickly. Changes in inspection periods and higher confidence would also result in higher utilisation of aircraft. Despite these considerable advantages, acoustic emission is not yet an accepted method for the examination of aircraft structures, mainly because of problems with background noise. The difficulties met when examining riveted aluminium plates (which constitute the majority of aircraft frames) are considerable. Work in this area is still at an early stage and so far only three major publications have appeared, although research is proceeding at a vigorous pace in both the United States and the United Kingdom. The major difficulty that has been encountered is that of background noise, but there is a difference in opinion expressed between the papers of Bailey, Hamilton and Pless (1976) and that of Dingwall (1977).

The first paper describes both in-flight tests and fatigue testing of a production-size wing. The good correlation of counts with crack lengths during crack initiation and even better correlation during crack propagation obtained in the fatigue tests are shown in figure 7.1. It will also

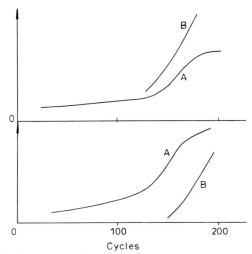

Figure 7.1   Correlation of acoustic emission counts, A, with crack length,
B, for an aircraft wing.

be seen that the cumulative number of events is very low compared with some other acoustic emission applications, except during the latter phases of the crack growth period, when up to 2×10^5 counts were observed from the beginning of the test. The authors were able to make a detailed study of the crack by exposing the crack surfaces to a fractographic process. Analysis of events per cycle during each of the load cycles midway through the test, showed that some emission was observed even when the crack did not seem to be growing visibly. This is especially so during load cycles 36 to 39 (out of a total of some 106). Considerable evidence is presented in the paper to show that acoustic emission can detect concealed crack growth — that is when the material is being cracked but no effect is visible on the surface.

The in-flight tests showed that the level of background noise (above 500 kHz) in flight is low enough to allow detection of signals from very small fatigue crack extensions in 7075 T6511 aluminium alloy. In these tests the transducers were single-ended with a resonance frequency response in the 700–800 kHz range, bonded to the surface with an acrylic adhesive fast-curing at room temperature. The equipment had a high-pass frequency filter with a cut-off at 350 kHz, falling above that at 48 dB per octave. Amplification within the channel was approximately 70 dB.

In the fatigue tests the test area was monitored continuously, the transducers being located in a centred triangle with the centre point within 2·5 cm of the predicted crack path. Growth was monitored by three methods:

(1) stopping load cycling and measuring the crack lengths observed by eye;
(2) continuously drilling a surface from a remote location through a system of telescopes and mirrors;
(3) constant surveillance of the acoustic emission responses with interpretation by a skilled engineer.

Video tape recordings were also made of the test area.

The contrasting work by Dingwall (1977) was carried out at the Royal Aircraft Establishment, Farnborough, using equipment specially designed for this type of investigation. Dingwall and his co-workers used a special system employing guard ring transducers. The main purpose of the test was to establish what noise levels could be expected when working on this type of structure and to see whether growing fatigue cracks could be detected against background noise using the guard ring system designed by Surrey University.

As soon as the tests were started it immediately became obvious that the aircraft was exceedingly noisy, both from the guard ring instrumentation results and also from simple observation. For example, during 10 minutes 70 000 counts were obtained on a signal channel, with no less than 1·52 million counts on the guard channel. Signals of 70 mV peak-to-peak were seen. This has to be taken in conjunction with the fact that signal attenuation in the structure (that is attenuation of a signal passing along the aluminium sheets) was of the region of 30 dB m^{-1} with an attenuation of about 25 dB across a simple joint. Dingwall was most discouraged by the level of noise in this structure and it is evident that guard rings or equivalent technology will have to be extensively used in aircraft applications. One alternative is the method of selective simultaneous signal detection described in an earlier chapter.

Cracks were observed by Dingwall, but the transducers had to be placed within one or two inches of a growing crack, which was detected despite background noise. Dingwall was confident that growing cracks can be detected in most aluminium alloys and can give an indication of the number of counts to be expected per centimetre of crack growth.

The system used half-wave rectification, with the overall bandwidth set at 70 kHz to 1 MHz (as measured at the 3 dB points). The transducers were of approximately the same sensitivity. A four-level amplitude sorter had the most sensitive trigger set at 96 dB, relative to 1·7 V with the others at 20, 40 and 60 dB amplitude. The guard ring is an extremely important feature of this system. Two channels were used in the amplitude sorter: the first being the actual signal channel, the second having three transducers connected to it, arranged in a ring around the signal channel. The equipment then operated as described in section 2.2, confining the field of surveillance of the signal transducer to a region around it of half the distance between guard and signal transducers. This system was checked at the beginning of the aircraft trial with the signal and guard transducers at 120° and 40 cm from the signal transducer. System testing was carried out with another transducer driven from a pulse generator. With this pulser outside the guard ring, no counts were obtained on the signal channel while the fatigue test was not running.

The difference between Dingwall's results and those of Bailey, Hamilton and Pless can be explained by the fact that their transducers cut off sharply below 500 kHz, compared to the broad-band system used by Dingwall. Also the rig used by Bailey, Hamilton and Pless was somewhat quiet compared to the noisy RAE rig. It could well be that this indicates some hope for in-flight testing, where background noise due to hydraulic rams would, of course, be absent.

7.2   Helicopter rotor systems

The Bell Helicopter Company has carried out laboratory and in-flight tests of a system designed to monitor rotor blades, based on Dunegan equipment but with a special modification due to the Bell Company itself. These results showed that an acoustic emission method using rise time discrimination gave indications of impending failure during a fatigue test on a blade section, with failure preceded by an increase in the acoustic emission count rate.

One of the potential difficulties in implementing the acoustic emission technique in rotor systems is the high attenuation of the acoustic signal in the honeycomb-supported section of the blade. The problem is alleviated to some extent by the fact that acoustic signals are transmitted with much greater efficiency through the doubler, leading edge, and trailing edge of the blade. These blade sections can serve to transmit signals of failures in the honeycomb but if far-removed from one of these high transmission efficiency regions, sources would be difficult to detect. Unfortunately transducers cannot be located in the hub area to detect failures in the blade section because of the high attenuation of acoustic signals across rotor system indicators. This procedure is also prohibited by the relatively high attenuation values in the blade itself.

In these tests a discriminator measured the rise time of the acoustic pulse envelope and rejected pulses having rise times less than a selected value. The range of rise times covered with this instrument varies from 0 to 10 V ms^{-1}. The discrimination between acoustic emissions and extraneous noise depends on the extraneous noise pulses having slower rise times than the acoustic emission pulses. A spatial gating method also used to try to reduce background noises was essentially the same as that described for offshore applications in the preceding chapter. This was of use unless the crack area was found to be in the area of the retention hub, the spindle or the grip of the blade when extraneous noise precluded the use of the spatial filtering method. Frequency bandpass filtering was also found to be ineffective for background noise discrimination because extraneous noise and fatigue-produced acoustic emissions fell into the same frequency band (100–300 kHz).

References

Bailey C D, Hamilton J M and Pless W M 1976 *Non-destr. Test. Int.* **9** 298

Dingwall P 1977 *Paper presented to EWGAE, Rome, October 1977*

8 The Application of Acoustic Emission to Fibre-reinforced Materials and to Concrete

8.1 The use of acoustic emission for testing fibre-reinforced materials

There has been a significant growth in the use of fibre-reinforced materials in engineering applications in the last few years. Glass-fibre-reinforced plastics (GFRP) are now widely used in boat construction, for vessels made to hold chemicals and in the manufacture of pressure vessels designed to withstand several hundreds of atmospheres. Carbon-fibre composites in which the matrix material is either polymeric or carbon are now used in transport applications such as military aircraft. There are also new applications for all these materials in automobile manufacturing, especially in sports cars, truck leaf springs and propellor shafts. The military application of carbon-fibre-based material has led to some signficant work on the use of acoustic emission: in contrast, GFRP material has received little attention from acoustic emission specialists.

Bunsell (1977) has tested pressurised rings and pressure vessels made of carbon-fibre-reinforced plastics (CFRP), showing that (figure 8.1) at about one-third of final failure stress there is a significant onset of total acoustic emission counts. In this type of material, failure occurs in three stages: first, cracks appear in the matrix, then the fibre–matrix band is damaged, and finally the fibres (much stronger than the matrix) fail. Figure 8.2 shows how CFRP is acoustically quiet on reloading a second time in a cyclic test until approximately 93% of the previous maximum load, with most of the emissions occurring during the first cycle. These figures show that the acoustic emission recorded during first cycle represents irreversible damage. Bunsell also showed that if a CFRP specimen is cycled to a maximum load within 4% of its fracture stress, it will eventually fail through the accumulation of damage, all of which could

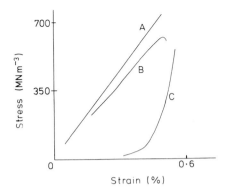

Figure 8.1 Acoustic emission counts for CFRP materials: A, CFRP; B, uncured CFRP; C, total counts (schematic).

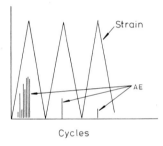

Figure 8.2 Acoustic emission from CFRP in fatigue tests (schematic).

be monitored using acoustic emission techniques. The specimen would fail before emissions had finished completely but there is no simple relationship between cyclic stress level and either the total number of counts to failure or the number of cycles for the finishing of emissions. No samples were found to have failed after they had ceased emitting. Thus these important results indicate that samples of CFRP can be tested for structural integrity using acoustic emission.

Pressure was also applied hydraulically to CFRP rings, confirming the results obtained with unidirectional specimens. Likewise, a test on a pressure vessel operating at $10.5\,\text{MN m}^{-2}$ gave similar results. The conclusion can be drawn from this work that a useful proof test for CFRP would be provided by loading the material and holding on a constant load for ten or twenty minutes; if the structure is then unloaded and reloaded, the load at which acoustic emission recurs can be considered a safe maximum working load. Even without monitoring the emissions, it appears that the type of CFRP structures tested in these studies can be used with competence at loads up to 93% of the previous maximum applied load.

Military applications frequently call for the use of carbon-fibre-reinforced carbon composites, and work at the UK Ministry of Defence (see Fry 1976) has shown acoustic emission to be useful for studying failure mechanisms in carbon–carbon composites and their precursors. Microstructural differences in the range of materials investigated can also enable acoustic emission to distinguish composites manufactured by different process routes. It has also been shown that the higher the stress at which the first significant energy peak occurs, the greater the failure stress of the composite. The results obtained during cyclic testing of carbon–carbon composites suggest an application as a quality assurance test technique. An upper limit can be set to the total number of counts monitored after the first complete cycle, leading to a way in which low-strength composites can be monitored. Figure 8.3 shows how the ultimate flexural strength varies with the stress at which the first significant emission amplitude peak occurs. Figure 8.4 shows the relationship between the ultimate flexural strength of unidirectional carbon composites and the number of acoustic emissions monitored, excluding

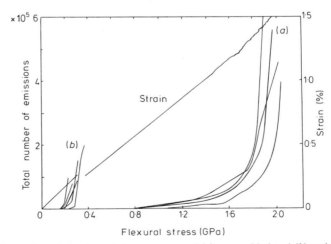

Figure 8.3 Acoustic emission responses of (*a*) as-moulded and (*b*) carbonised HTU–FPN8 composites and typical stress–strain diagram (after Fry 1976; Crown Copyright reserved).

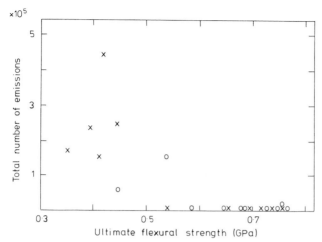

Figure 8.4 Relationship between the ultimate flexural strength of sandwich structure composites and the number of acoustic emissions monitored, excluding the first cycle, during flexural cycling from 0 to 0·2 GPa. The total number of cycles was ten. ×, 11 unidirectional layers; ○, 4 unidirectional layers (after Fry 1976; Crown Copyright reserved).

the first cycle, during flexural cycling over 15 cycles. It can be seen that the 'good material' has a low number of cycles whereas the poor material has copious emissions. Other Ministry of Defence work in this area has included a technique for detecting acoustic emission devised by Dean and Kerridge (1976). In the USA Green *et al* (1964) have used acoustic methods to carry out quality assurance procedures on Polaris rocket chambers.

Lloyd-Graham has described the elaborate AEMPA system developed by the Rockwell Corporation for testing carbon-fibre epoxy pressure vessels. A wide-band (10 kHz to 1·8 MHz) system is coupled into a computer capable of assigning 23 parameters to a set of acoustic emission signals, including mean amplitude, rise times, and impulse frequency (in terms of 8 octaves). He found that the parameter

$$P_f = \text{(amplitude at 56 kHz)/(amplitude at 556 kHz)}$$

allowed the nature of a crack to be determined via a set of characteristic P_f–time plots. The P_f–time curve for a delamination, for example, is quite different from that for a set of fracturing fibres.

8.1.1 Tests at the Fulmer Research Institute on fibre-reinforced materials

An extensive series of tests has been performed on CFRP/Al honeycomb samples at the Fulmer Research Institute (W E Duckworth, private communication) to determine the effectiveness of various NDT methods for locating faults. In addition to the NDT work, a programme of destructive testing was undertaken, involving four-point bend flexural testing of specimens containing deliberately introduced defects. These tests were monitored for acoustic emission activity. Also, a programme of tests on deliberately defective CFRP/Al honeycomb dishes was performed, using vibration methods and acoustic emission monitoring.

The defects studied included solvent contamination, overlap joints in the CFRP, contamination in the laminated skins and in the core–skin bond, crushed and damaged aluminium cores and debonding between skin and core. Two types of CFRP skins were used, both formed using the pre-impregnation method using indirectional fibres: however, one set of skins consisted of four-ply laminates in the 0,90,90,0 type of lay-up and the other of a three-ply 0,60,120 type. An amplitude sorter was used to differentiate between various types of event, e.g. resin cracking, fibre breakage, pull out and so on. The main results of the monitoring were as follows.

(1)   The acoustic emission characteristics of all nominally perfect specimens were similar to each other but distinctly different from all specimens containing defects. It was typical of the nominally perfect specimens that no significant emissions were detected until immediately before failure. The behaviour of specimens containing defects was consistent within groups containing the same defect type but different between groups. All defect types caused the generation of emissions early in the loading history and the distinction between types was in terms of the energy of the emissions and the general specimen strain at which they were detected.

(2)   The acoustic emissions which were detected at low load levels corresponded to the growth of resin cracks in interlaminar planes or skin–core boundaries. These cracks modified the deformation behaviour and caused early fibre failure.

(3)   Acoustic emissions which occurred at low loads tended to be generated in short bursts which corresponded to bursts of crack growth. Crack growth steps caused redistribution of strain con-

centration which led to temporary arrests until higher load levels were imposed. Results in vibration tests suggest that cracks stabilise at particular stress levels and further cycling does not lead to further growth in the short term. This does not give an indication of long-term stability however.

(4) Testing of a specimen cut at 45° to the fibre directions confirmed that the low-energy emissions were due to resin cracking rather than fibre failure.

The vibration tests were inconclusive in locating and identifying defects that were growing during the tests. However (as in the case of the original work on the Marots Antenna) significant emissions were detected and in particular one sample which failed catastrophically gave emissions from very early in the testing, i.e. at low '*g*' levels.

A great deal of work remains to be done to provide a reliable indication, of fault location and type, but the differences in acoustic emission signature from the various defect types do indicate a potential way of differentiating between them.

8.1.2 Testing bonded joints

Associated with the use of acoustic emission to test fibre-reinforced structures is its use for characterising adhesive joint failure. This subject is still at a very early stage of development and only a few workers have studied it in depth (notably Beale, Pollock, Curtis and Hill). Hill, for example, has tested a number of lap joints using Redux 775RN vinyl–phenolic adhesive reinforced with PVC mesh and also an epoxy–phenolic adhesive. Various curing times and temperatures were tested in a very simple laboratory arrangement. Acoustic emissions were obtained and Hill concluded that they do reveal some of the detailed processes of lap-joint fracture not evident using normal testing techniques. For example it appears that the strength of one of the epoxy joints depends on the voids blunting the peeling crack, which is initiated relatively easily. Environmental conditions can reduce the strength of adhesive joints and acoustic emission measurements will give warning of degra-dation of joint strength, provided an adhesive which is only slowly dependent on temperature is used. Deliberate use of high stress con-centrations and lower operating loads may be needed.

The Kaiser effect is not shown at low loads but agreement with the law first propounded by Kaiser improves at loading nearer to joint failure. Hill has shown that the most suitable criterion for joint failure appears to be that of reaching the previous count rate, although onset of emission after a previous load cycle may be suitable when using loads well below the fracture load. In other words Hill suggests the same criterion for failure as used for epoxy and other reinforced materials and he also points out that not every adhesive is a good generator of acoustic emission.

This subject is obviously worthy of much greater detailed investigation than has been afforded by the small number of publications so far available, but it must be said that the determination of adhesive joint failure is not at such an advanced stage as is the case with welded joints.

8.2   The use of acoustic emission for testing concrete

Many workers have naturally turned their attention to the use of acoustic emission for the non-destructive testing of concrete structures because of the widespread use of this material in the construction industry. STP 505 contains several references to tests on concrete structures, without any promise of successful application. Recently, however, an acoustic emission method has emerged which should prove to be of value in testing concrete structures made from high-alumina cement (HAC). In this material, which is quite different from conventional Portland cement, the final strength is achieved very rapidly, in about one day (compared with several days for conventional Portland cement). Some high-alumina beams, after having been in position for several years, have shown signs of cracking. There have even been cases of failure leading to injury or loss of life. Consequently the non-destructive testing of HAC reinforced beams is a matter of importance.

Concrete is very different from metallic materials. It is essentially composite rather than alloy in nature, consisting of gravel, cement and, in many cases, reinforcing steel bars. The situation is further complicated by the fact that the steel bars can be pre-tensioned. Again, unlike metals, the material can change its strength and composition markedly during its service life. Tests on concrete cubes and pre-stressed beams (in both cases made from HAC) had been carried out by the Fulmer Research Institute and by Acoustic Emission Consultants Ltd in cooperation with the British Gas Corporation (Arrington and Evans, 1977). These results, although of a somewhat preliminary nature at this stage, give a hope

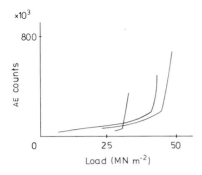

Figure 8.5   Emissions from HAC concrete.

that acoustic emission could be applied to the testing of HAC reinforced beams.

Figure 8.5 shows emissions from HAC concrete cubes tested for the British Gas Corporation. Four phases can be distinguished.

(1)   In the first phase there is an initial loading noise caused by the first increments of loading, mainly due to the crushing of debris from the previous test.

(2)   At small values of the load the initial emission rate rises linearly with load. In this region there is very little damage and the main source of the emission activity is thought to be compaction of the concrete matrix.

(3)   At about 60% of the load, emission starts to rise exponentially indicating departure from elastic behaviour. This region corresponds to the onset of cracking and to the situation where lateral strains in the cube test become measurable.

(4)   It can be seen from figure 8.5 that at about 90% of the failure load the emission rises very rapidly, due to the onset of large cracks in the test piece.

The importance of these graphs is evident. They show that at approximately 60% of the breaking load, a detectable change in the emission pattern occurs, and at approximately 90% of the breaking load a very rapid increase in emission takes place.

In the USA publications have been sparse. Nielsen and Griffin (1977) discuss the Kaiser effect in plain concrete and a recent publication by the American Concrete Institute (NDT *Methods No. 9: Testing Hardened*

Concrete) also briefly discusses the use of acoustic emission for testing concrete.

8.2.1   Acoustic emission testing of high-alumina cement concrete beams at the Fulmer Research Institute

FRI staff have studied the acoustic emission behaviour of ×5 HAC concrete beams during four-point flexural deformation, and have proposed as a result of this work an NDT procedure for HAC beams which would allow estimation of the remaining strength of beams and could determine whether or not they were cracked.

The Fulmer workers aimed to find:

(1)   What were the ultimate flexural loads and deflections for the beams, and how did the acoustic emission rate vary with load during simple flexure testing?

(2)   When loads applied to the beams in flexure approached failure loads, could creeps or imminent failure in the beams be detected by acoustic emission techniques?

(3)   When the beams were subjected to simulated service loads, could their extra load-bearing capacity be assessed from acoustic activity in an overload test?

(4)   Did the Kaiser effect operate to give a reliable indication of whether or not beam properties had deteriorated between tests?

(5)   At what proportion of the ultimate load did the Kaiser effect break down because of creeps or progressive damage under a fixed load?

The beams tested in this work, supplied by the Building Research Association, were tested in a four-point bend rig with spans of 3·05 m and 1·017 m between support and loading rollers respectively. Loads were applied with a Mand 650 kN servohydraulic testing machine. The acoustic emission equipment comprised various Dunegan Endevco modules, including an energy processor, and the transducer was an Endevco D140B resonant differential device.

The results of the work showed primarily that many more tests than were possible in this project would have to be carried out to give a reliable guide to the acoustic emission behaviour of a material as inherently variable as HAC concrete. However, some quantitative observations could be made.

Eight of the eleven beams tested showed acoustic emission activity at 62·1 (±3·4)% of their ultimate load and 87·1 (±2·4)% of their tensile cracking load. The Kaiser effect occurred until the beams cracked. i.e. until macroscopic damage occurred and the beam compliance changed. After cracking the Kaiser effect was no longer observed. If a beam under load was given a small overload and displayed acoustic emission activity, it was approaching, or beyond, its cracking load. If, when the overload was removed, further activity was detected the beam was cracked; if there was no activity the beam was intact. If a slight overload causes activity and the overload is held constant, the emission will show a decreasing rate with respect to time.

These summarised results led to some tentative suggestions for acoustic emission monitoring of concrete beams.

(1)　If no emission activity was detected during loading of a beam, then the applied load was less than 62·1 (±3·4)% of the ultimate strength of the beam. If acoustic emission did occur at a level judged to be significant, the load had exceeded this percentage strength. If on unloading, further emission occurred, the beam was cracked; if no emission occurred, the beam was not cracked. If the beam was cracked the quantity of emissions on unloading may indicate the margin by which the cracking load had been exceeded.

(2)　A beam under service load would give similar behaviour on application of an overload, the condition of the beam being indicated by the presence or absence of acoustic emission activity during loading or unloading.

(3)　If a beam showed no activity under an overload, but, at a later date, did show activity under the same overload its strength must have decreased in the interval between tests.

8.2.2　Acoustic emission testing of ordinary Portland cement pre-stressed concrete beams

At the Fulmer Research Institute eight ordinary Portland cement (OPC) pre-stressed concrete beams of four different designs have also been tested to failure in one- or two-point flexural loading. Emissions were first generated in every case when a load equivalent to 89 (±6)% of the load required to cause cracking was applied. Cracking was invariably associated with the limit of proportionality of the beams. When uncracked beams were unloaded no emissions were detected, and during reloading no emissions were detected until the previous maximum load had been exceeded. When cracked beams were unloaded a high level of emission occurred and during reloading a low level of emission occurred until the previous maximum load had been exceeded.

The results from the OPC beams were similar in all respects to those found with the HAC beams and, in fact, a common test procedure could be adopted for all types of beam.

8.2.3   The use of acoustic emission to monitor the condition of concrete structures

Pre-stressed HAC and OPC concrete beams have been monitored at the Fulmer Research Institute for acoustic emission during flexural testing to failure. These experiments demonstrated that acoustic emission actively commences at 62·0 (±3·5)% of the ultimate moment for HAC beams and 59·6 (±4·5)% of the ultimate moment for OPC beams. Monitoring of acoustic emission could also be used to distinguish between cracked and uncracked beams, since cracked beams generated emissions during unloading, whilst uncracked beams did not. Cracking of good beams occurred when they were subjected to a load equal to approximately 69·0% of ultimate for HAC beams and 71·0% of ultimate for OPC beams. It was further observed that if a load, less than that required to cause beam cracking but exceeding that required to cause emissions, was applied to a sound beam, removed and then reapplied, no emissions were observed in the second cycle unless the previous maximum load had been exceeded. This effect was not observed if the beam was cracked initially or if the cracking load was exceeded in the first loading.

These observations formed the basis of a method suggested for assessing the condition of in-service beams. The method involved the application of an overload to a beam in service. The implication was that if no emissions were detected, the overload had raised the stress in the beam to less than 59·6 or 62·0% of its ultimate, depending on the type of beam. If activity did occur, and it was judged to be significant, then the beam had been subjected to more than 59·6 or 62·0% of ultimate stress. If on unloading further emissions were detected, the beam would be cracked; if on unloading no emissions occurred, the load on the beam had taken it to between 59·6 and 71·0% or 62·0 and 69·0% of its ultimate, depending on beam type. If emissions did occur during unloading, the number of emissions would give an indication of the amount by which the cracking load had been exceeded and enable a judgement to be made as to whether or not a further examination was justified. If a beam showed no activity under the influence of an overload on one date, but at a later date did show activity under the same overload, this would be an indication that its strength had decreased between tests. The method could, therefore, be used for monitoring structures over a period of time.

Possible objections to testing beams in service include the suggestion that emission from sources unconnected with the beam cracking would be detected. Such sources could include movement between a beam and its points of support, ambient vibrations in the test vicinity, relative movement between beams and beam separators or between beams and a surface screed if a floor is considered. Consequently, a model floor has been built in the Civil Engineering Laboratory at the Central London Polytechnic and monitored for acoustic emission during a series of loading cycles at progressively increasing loads until it fractured. Six type X7–T9 OPC concrete beams of length 16 ft were supported on walls constructed from concrete blocks. The distance between the points of support of the beams was 15 ft and their separation was 2 ft between centres. The spaces between beams were filled with blocks (pots) of a cellular structure. The pots were made using a lightweight 'breeze' type material. The top surface of the beams and pots was covered with a 2 inch thick concrete screed. Several four inch square pockets were left through the screed to allow the positioning of transducers directly on beams.

Provision for loading the floor was made by the positioning of steel girders on top of the screed parallel to the supporting walls. The loading girders were 'acoustically insulated' from the floor by rubber mats on their lower surfaces and they were placed at 18 inches on either side of the centre line. Loading of each girder was through two steel cables connecting it to two hydraulic jacks which were situated under the laboratory strong floor against which they exerted their load. The loading cables thus passed through the test floor and the supporting floor. A load cell was connected between one loading cable and its steel girder. Since the floor hydraulic jacks were connected in parallel to the same pump, it was assumed that they were each supplying the same force.

The first monitoring system used a simple transducer situated on top of one of the floor beams. The second system used a single active transducer and three additional guard transducers. The function of the guard transducers was to prevent the active transducer, which was situtated in the centre of the lower surface of one beam, from being activated by acoustic signals originating on the screed, at the screed–beam boundary, at the screed–'pot' boundaries or at the points of support of the beam. The first significant emissions occurred during loading when the limit of proportionality for the floor was exceeded. In subsequent load cycles an imperfect 'Kaiser effect' was observed when the previous maximum load was passed. No emissions were detected during unloading until the beams were visibly cracked. Once this state was developed, very high count rates occurred during the unloading stage of all subsequent cycles.

This type of test appears to be reliable, it is easy to carry out and there is no reason why it should not be applied to systems in service. Careful interpretation of results from a first test should enable judgement to be made as to whether or not the limit of proportionality has been exceeded. If it has, it should be possible to determine whether or not cracked beams are present and what maximum load the structure has previously been exposed to. If cracked beams are present, further examinations using other methods will be necessary.

The method may also be used as a monitoring method to observe the rate of degradation of strength of a structure. In tests applied at intervals, the response to a fixed load, or the load required to give a fixed response, may be measured. The rate of change of these loads indicates the rate of degradation of structural strength.

References

Arrington M and Evans B M 1977 *Non-destr. Test. Int.* **10** 81

Bunsell A R 1977 *Non-destr. Test. Int.* **10** 21

Chabowski A J, Bowyer W, Cook J M and Peters C T 1977 *Proc. 8th World Conf. on Non-destructive Testing, Cannes 1977*

Dean D S and Kerridge L A 1976 *Non-destr. Test. Int.* **9** 233

Fry M J 1976 *AWRE Report No.* 42/76

Green A T, Lockmann C S and Steele R K 1964 *Plastics* **28** 137

Nielsen J and Griffin D F 1977 *J. Test. Eval.* **5** 476

Index

117